T5-BPZ-344

The "Global Negotiation" and Beyond

Toward North-South Accommodation in the 1980s

Edited by Roger D. Hansen

Lyndon B. Johnson School of Public Affairs
The University of Texas at Austin
1981

382
G 562

Library of Congress Catalog Card No.: 81-83959
ISBN: 0-89940-004-3

© *1981* by the Board of Regents
The University of Texas at Austin
Printed in the United States of America
Designed by Kathi Branson

82-660

The "Global Negotiation" and Beyond

To Gladys and Sid

Contents

Acknowledgements

While many people assisted in the preparations for the 1981 Slick Professorship Symposium, there are several whose efforts cannot go unmentioned. First, Dean Elspeth Rostow was a constant source of support and advice in the months preceding the symposium. For those efforts and for her typically warm and gracious hosting of the event I am deeply grateful.

Second, I should like to thank Professor Sidney Weintraub for so willingly and helpfully sharing with me his experiences relating to various previous symposia and his views concerning the most productive structuring of this one.

Third, I wish to extend my deepest thanks to all the participants who came to Austin for the event. Each overloaded an already crowded professional schedule to join in this two-day exchange, and I am deeply appreciative of this extra effort on their parts. As the symposium discussions illustrate so well, there was nothing *pro forma* about their participation; the spirit as well as the substance which distinguished the two-day exchange is an appropriate tribute to each and every one of them.

Finally, I take great pleasure in expressing my deepest gratitude to Cheryl McVay, who contributed in so many crucial ways to the preparation and coordination of the symposium—and to this publication itself—that any single title such as Symposium Coordinator necessarily understates the value of her assistance. I shall always be grateful for her professionalism and support during my nine-month visit to Austin.

The "Global Negotiation" and Beyond: Toward North-South Accommodation in the 1980s

March 27-28, 1981
East Campus Lecture Hall

A Symposium Sponsored by the Distinguished Visiting Tom Slick Professorship of World Peace at the Lyndon B. Johnson School of Public Affairs The University of Texas at Austin

**Lyndon B. Johnson School of Public Affairs
The University of Texas at Austin**

Foreword

For several years both rich and poor members of the world's developing countries have been calling for a year of global negotiations on the issue of international economic cooperation for development. The year 1981 has been chosen for these negotiations. However, the inability of the United States, Great Britain, and West Germany to reach a compromise with the developing countries on the terms of reference of these negotiations has delayed their opening.

Nevertheless, a conference devoted to an examination of the issues to be raised in the negotiations is timely for at least three reasons.

First, a Summit meeting of approximately twenty to twenty-five heads of state to examine a similar range of economic and political relations between developed and developing countries is already scheduled for October, to be hosted in Cancún by President López-Portillo.

Second, the issues to be examined in the "global negotiation" of 1981 remain to be dealt with regardless of the present deadlock over terms of reference for the negotiation itself.

Finally, the conference will serve as an appropriate setting in which to examine an agenda of issues which reflects economic and political conflicts of the 1960s and 1970s in the perspective of global economic and institutional needs of the new decade.

In analyzing both the old and the new problems challenging economic cooperation among developed and developing countries in the 1980s, the 1981 Slick Conference aims to reach some consensus on concrete steps required to diminish the level of North-South conflict over international economic and institutional issues which dominated discussions of the 1970s and to begin to work in closer harmony on the international economic challenges of the 1980s. And where consensus cannot be reached, a clear and dispassionate analysis of the causes of dissensus may well prove to be equally constructive. For in either case it will then be easier for all states, developed and developing, to make rational choices relating to international economic problems facing them with far better knowledge of all of the costs and benefits involved in their decisions.

Roger D. Hansen
1980-81 Distinguished Tom Slick Professor of World Peace

Friday, March 27

9:15 a.m.   WELCOME

Elspeth Rostow, Dean of the Lyndon B. Johnson School of Public Affairs

INTRODUCTION

Roger Hansen, Distinguished Visiting Tom Slick Professor of World Peace; Jacob Blaustein Professor of International Organization, Johns Hopkins School of Advanced International Studies, Washington, D.C.

9:30 a.m.
12:30 p.m.   **The Setting**

Chair **Donald Mills,** Jamaican Ambassador to the United Nations; former Chairman of the Group of 77

1. The Political and Economic Origins of the Global Negotiation of 1981

Mahbub ul Haq, Director of Policy Planning, World Bank

2. The "Global Negotiation": Agenda, Progress, and Problems

Joan Spero, U.S. Ambassador to Economic and Social Council, United Nations

3. Beyond the Official Agenda: Some Crucial Issues

Walt Rostow, Rex G. Baker Jr. Professor of Political Economy, The University of Texas at Austin

2:00 p.m.-
5:00 p.m.   **Energy and the North-South Impasse: Three Perspectives**

Chair **Sidney Weintraub,** Dean Rusk Professor of Public Affairs, Lyndon B. Johnson School of Public Affairs

1. A View from the Oil-Exporting Developing World

Jorge Eduardo Navarrete, Under Secretary for Economic Affairs, Mexican Ministry of Foreign Affairs

2. A View from the Non-Oil-Exporting Developing World

Donald Mills

3. A View from the North

John Foster, Senior Economic Adviser, Petro Canada

Saturday, March 28

9:30 a.m.-
12:30 p.m.   **The Bretton Woods Institutions: What Room for Accommodation?**

Chair   **Paul Streeten,** Director of Center for Asian Studies, Boston University

1. Overview: Altering Tasks, Changing Institutions

Mahbub ul Haq

2. Finance and Development

Paul Streeten

3. Trade and Development

Gerald Helleiner, Professor of Political Economy, University of Toronto

2:00 p.m.-
4:30 p.m.   **Can a "Mutual Interest" Agenda Evolve During the Global Negotiation?**

Chair   **John Sewell,** President, Overseas Development Council

1. An Economic Perspective

Nícolas Barletta, Vice President of Latin American and Caribbean Region, World Bank

2. A Political Perspective

Edward Hamilton, President, Hamilton-Rabinovitz & Szanton, Inc.; former Executive Director, Pearson Commission

Introduction
by Roger D. Hansen

The six-month history of this year's Slick Symposium topic — the "Global Negotiation" — is indicative of both the complexity of the issues involved in North-South relations and the continuing persistence with which they are raised.

The topic was chosen in early September of 1980 when it still appeared certain that six to nine months of 1981 would indeed witness the so-called "Global Negotiation." For several years both rich and poor members of the world's developing countries have been calling for a major negotiation on the issue of international economic cooperation for development. At the beginning of 1980 the process of preparation for this negotiation began at the United Nations in New York, and it was intended that all agreements necessary to open the talks in 1981 would be completed at the meeting of the Eleventh Special Session of the UN General Assembly in August of 1980.

But agreement was not reached. In late September the United States, joined by Great Britain and West Germany, rejected the compromise package accepted by all other UN member states. With the close of the Eleventh Special Session, the problem was passed back to the UN General Assembly for further attempts to reach a consensus.

As of April 1981, consensus had not been reached, and the "Global Negotiation" had therefore not begun. To understand the origins and development of the conflict between the United States (supported by two Northern allies) and the developing countries that has thus far stalled the beginnings of any "Global Negotiation," the reader is referred to the first session of the symposium, particularly the opening remarks of former Ambassador Joan Spero, who represented the United States at most of the negotiating sessions on this subject held during 1980 and early 1981 in New York. A reading of this session of the symposium provides very interesting insights into the complexity of the issues involved in launching this particular negotiation. The reader will quickly note that the issues of *process* are often as important as those of *substance*, since procedural factors (e.g., voting rules) can often have a major impact on substantive issues. Procedural decisions not only heavily influence what the substantive issues for negotiation will be, but also the outcomes of the negotiation. For example, the greater the role of the UN General

1

Assembly as a coordinating body and court of last resort in the negotiations, the greater the prospects that the world's developing countries (i.e., the "South," in current jargon) will control the major substantive outcomes of the negotiation due to their overwhelming numerical majority in the UN General Assembly.

If the failure thus far to reach agreement on the "Global Negotiation" reveals many of the complexities in the current North-South conflict, the surprisingly rapid evolution of plans to hold a meeting of some twenty-three heads of states from both developed and developing countries in Cancún, Mexico, in October of 1981 underscores the persistency of the South — with a considerable and growing amount of Northern support — in raising the set of issues to be considered in the "Global Negotiation." The reader is referred to Jorge Navarrete's remarks in the final session of the symposium for an overview of the items likely to be discussed at the Cancún summit. While details are not yet entirely resolved, it appears that the major issues for discussion will include food security and agricultural development; trade rules and industrial development; energy, with particular reference to the energy-deficient developing countries; and international financial and monetary problems, including the long-term problem of adequate resource transfers to the world's least developed countries.

It now appears that the three countries which are still blocking the opening of the "Global Negotiation" via the United Nations will all attend the Cancún summit, where they will discuss many if not all of the same substantive issues they would have to confront in the former negotiating context. The setting and procedural rules will of course be quite different; furthermore, as noted in the final session of the symposium, there will be no attempt to negotiate specifics at the Cancún summit. The point worth noting is not that developing countries, stymied in one venue, have successfully created an equally promising one in which to press for the same goals. This would be an incorrect reading of the comparison between what is being attempted by the developing countries via the "Global Negotiation" under UN auspices and what is being attempted at Cancún. The relevant point is that, through a successful call for the Cancún summit, the developing countries are able to keep their concerns actively before the international community in one form or another.

They have demonstrated this ability throughout the past decade, most particularly since the Sixth Special Session of the UN General Assembly, held in April and May of 1974, which culminated in the call for a New International Economic Order (NIEO). In the succeeding years the so-called NIEO demands of the developing countries, discussed throughout the symposium, have been pressed upon the developed countries at all meetings of the UN, the United Nations Conference on Trade and Development (UNCTAD), the United Nations Industrial Development Organization (UNIDO), the Third UN Conference on the Law of the Sea (LOS III), and a legion of other forums.

If the success of the OPEC countries in challenging the North in 1973–74 with embargoes and a quadrupling of crude oil prices ignited this veritable explosion of developing country diplomatic activity, more fundamental explanations must be traced to a long and complex institutional and psychological history. The process of Southern policy coordination over a broad range of diplomatic activities increased rapidly during the 1970s, in large part reflecting the strengthened institutional capabilities of the two organizations that represent Southern countries when they act as a unit in international relations (and create the aggregate North-South conflict), namely the so-called Group of 77 (G-77) and the Nonaligned Movement (NAM). The former organization's membership now includes close to 120 countries, while the latter embraces 93. Both are increasingly heterogeneous, with many tensions and the potential for disintegration always present. But these are held in check by a shared perspective on present North-South relations and by shared goals in the global arena that have very deep historical roots generally overlooked by Northern policymakers.

It is the shared perception of global inequity that can link over 120 "developing countries" with annual per capita incomes ranging from $200 to $10,000. Built on the historical evolution of North-South relations in both precolonial and postcolonial periods, this perception provides the cement that binds the otherwise disparate and potentially discordant membership of the Southern bloc. It has been the most significant driving force behind the activities of both the Group of 77 and the Nonaligned Movement since the inception of these groups close to twenty years ago. While the Nonaligned Movement was initially inspired by the idea of avoiding involvement in the Cold War, it soon became the preeminent international platform for attacks on colonialism, apartheid, "neocolonialism," and all other forms of Northern or white domination, real or imagined. Failure to comprehend the depth of this widely shared Southern perception was primarily responsible for endless inaccurate Northern predictions throughout the 1970s that an inevitable appreciation of divergent national interests would quickly dissolve any effort of developing countries to institutionalize "Southern unity."

For readers who come to the subject without extensive familiarity with the North-South conflict of the 1970s, the major NIEO demands deserve a brief summarization. They can be grouped into five areas: international trade, economic aid, foreign investment (direct and indirect), technology transfer, and the international monetary system.

In the trade field the major demands have been for (1) various forms of international commodity agreements that would raise and stabilize the price of raw materials exported by less developed counties; (2) a "common fund" that would provide a financial umbrella for these commodity agreements; (3) nonreciprocal reductions in developed-country barriers to developing-country exports of processed raw materials, semimanufactures, and manufactured goods; (4) expanded generalized trade preferences for the developing countries to better enable them to compete with industrial production in the markets of the North; and (5) better-financed domestic adjustment assistance programs in the North which, by easing transitional pains accompanying the restructuring of Northern economies, will facilitate imports of Southern manufactured (and some agricultural) goods. In sum, at the aggregate level the South has thus far sought increased relative prices for its raw material exports to the North as well as an increase in the volume of raw material and manufactured exports.

In the field of aid, Southern demands have thus far included the following: (1) that the developed countries meet the aid targets most of them agreed to in the International Development Strategy for the Second UN Development Decade (a minimum of 0.7 percent of GNP in the form of official development assistance and 1.0 percent of GNP including private capital flows to the South); (2) that the North increase its financial commitments to all those emergency funds created in response to the food and oil price increases of 1973-75; and (3) that the North be prepared to renegotiate the terms of debt repayment for those Southern countries experiencing serious balance-of-payments problems, including the possibility of significant debt cancellation.

In the area of foreign investment, one again finds several distinct Southern goals. One is greater access to international capital markets for Southern countries. A second is the elimination of traditional international legal restraints on the expropriation of foreign direct investments, including the putative requirement of "full, prompt, and effective" compensation. A third Southern goal has been to engage Northern governmental assistance in policing the activities of Northern multinational corporations for the general purpose of increasing the level of economic benefits going to host countries. Such increased benefits might take various forms: for example, greater North-South capital and technology flows, lower charges on technology transfer, less protection for existing (and overwhelmingly Northern) patent rights, and increased developing-country exports to the North of products of foreign corporations.

In the field of technology transfer, Southern demands and objectives include Northern financing earmarked for the creation, expansion, and modernization of Southern scientific and technological institutions; Northern governmental support in "persuading" multinational corporations to adapt their technology to host-country development needs; and international support for changes in patent laws and other measures that will lower the cost of technology transfers to the South.

Finally, with regard to the international monetary system, the developing countries in the aggregate continue to demand (1) a greater voice in the reform and management of that system; (2) more "automaticity" in access to those sources of reserves now available through IMF "windows" and less IMF surveillance over the use of their funds; and (3) the establishment of an international reserve creation process that automatically places increasing amounts of international reserves in the hands of developing countries.

The endless meetings between developed and developing countries since 1974 have made only the most marginal progress in achieving agreement on this broad range of demands/requests/proposals. Some have been dropped and others have been added, but the lack of affirmative results remains constant.

The purposes of the 1981 Slick Symposium were four. The first was to analyze the origins and present state of the North-South stalemate. The second was to examine some of the most pressing problems that are presently a part of the ongoing North-South discussions. A third was to underscore some significant international economic issues which have thus far remained at best marginal to the North-South "dialogue." And the fourth was to illuminate any and all prospects for constructive efforts to overcome the present deadlock in the next several years.

It was with these purposes in mind that symposium participants were chosen. Since the overreaching purpose was to explore the present deadlock for ways to move beyond it, only persons with an interest in finding "solutions" involving varying forms of cooperation between developed and developing countries were represented. As Professor Streeten noted in his closing remarks in Session III, this means (for better or worse) that certain perspectives were not represented at the symposium. Since he makes the point so well, it need not be developed here except to alert the reader that, despite the wide degree of disagreement to be found throughout the symposium, with a broader set of perspectives among the participants the lack of consensus would have been far greater. The point can best be summarized as follows: all symposium participants presumably shared a conviction that "mutually accommodative" solutions to the present North-South conflict *can* and *should* be found. Unrepresented were perspectives which held that negotiated accommodation was *impossible* and/or *undesirable.*

The symposium was divided into four sessions. In each session there were two or three oral presentations of major themes followed by general discussion among all thirteen participants. In each session the chairman opened with a brief statement linking the issues to be considered to the overall North-South problem and to the goals of the symposium. Additionally, each session was closed by the chairman's brief summary of the preceding discussion. The following comments on each session are offered in hopes that they will provide an appropriate and helpful introductory orientation.

Session I. The Setting

In this session Mahbub ul Haq's presentation bore little resemblance to the assigned topic for two reasons. First, as he himself noted, some prodding questions by Walt Rostow about the NIEO and its intellectual basis raised the evening before the symposium opened led Haq to offer a response at the very outset of the symposium. And having done so, he felt it more important to turn directly to a discussion of how the North and South might break out of their present stalemate. He did so by offering a seven-point program.

Joan Spero followed with an analysis of the major points of disagreement that led the United States to veto the compromise resolution of the Eleventh Special Session which would have led to the opening of the "Global Negotiation" in early 1981. In her presentation Spero offers an illuminating "inside" picture of the strengths and weaknesses of the negotiating process on North-South issues which takes place in the UN and related international organizations. A serious question raised by her analysis involves the degree to which the *process* of negotiation itself increases the seeming intractability of the issues at stake.

Finally, Walt Rostow offered an overview of the NIEO and its agenda in very sharp contrast with that delineated by Haq. In Rostow's view, the past and present intellectual constructs and concrete issues implicit and explicit in the NIEO demands for international economic reform were and are misspecified and misplaced. More important, in his view, they are doing a disservice to North-South relations by blocking joint efforts to tackle a "new agenda" of global problems in which both sides have vital interests. That new agenda is spelled out in his initial presentation, and includes a sharp focus on such issues as food, energy, and the environment.

Criticizing method as well as content, Rostow emphasized the need to downplay issues of principle and "global" bargaining by substituting a *pragmatic bargaining style* and a *regional approach* to problem-solving. Rostow's observations led to an exchange fundamental to the understanding of the North-South problem, an exchange to be repeated in several important respects in the final session of the symposium. Because of the importance of these two exchanges, they will be analyzed in some detail in the Postscript.

As in each session, the participant discussion which followed the initial presentations offers the reader a fascinatingly — and perhaps frustratingly — broad spectrum of views on the issues raised by the three speakers. No attempt will be made here to synthesize these "round table" reactions. To do so would only detract from their richness and their complexity; therefore all attempts at synthesis and integration are reserved for the Postscript.

Session II. Energy and the North-South Impasse: Three Perspectives

The three presentations of Session II underscored the complexity of the energy issue not only as a global challenge to all states but also as a lever in the North-South conflict. Jorge Navarrete opened the session by emphasizing the view that any "negotiation" on the issue of energy (or oil alone) would have to be a part of a broader "global" negotiation in which oil and energy issues were linked to other major issues of international economic reform that are perceived as necessary and desirable by the developing countries. This has been the publicly stated view of literally all the oil-exporting developing countries ever since 1974. At that time the North, in reaction to price increases and the oil embargo introduced by many of the OPEC states during the Yom Kippur war of 1973, attempted to isolate the OPEC states from the rest of their Southern diplomatic partners and force them into an energy "dialogue" with the OECD countries.*

Furthermore, this has been the position of all the non-oil-exporting developing countries as well, as indicated by Donald Mills's presentation which follows. These countries, hoping to use the "oil weapon" to force otherwise unattainable concessions from the North, have insisted and continue to insist that Northern desires to establish an ongoing "energy dialogue" with the OPEC states will be possible only if the North will simultaneously discuss and negotiate a host of related issues raised by the NIEO demands. Yet, as Mills spells out in illuminating detail, the relationship between the OPEC states and their oil-importing developing-country diplomatic partners is replete with its own kind of tensions and conflicts. In tracing the evolution of this crucial relationship within the Group of 77, Mills offers an insider's view of one of the most important elements of present Southern negotiating strengths — the OPEC "alliance" with other members of the Group of 77 — and encourages us to consider the future evolution of that alliance.

Viewing the energy problem from quite a different perspective, John Foster completely dismissed the idea of a "global bargain" on the issue of oil supplies and prices. His analysis of all the uncertainties and conflicting motives leading to this conclusion were echoed throughout the discussion period of the session, although opinion regarding that conclusion again differed widely.

---

*For a brief discussion of Northern strategy in this period, see Branislav Gosovic and John G. Ruggie, "On the Creation of a New International Economic Order," *International Organization* (Spring 1976).

Session III. The Bretton Woods Institutions: What Room for Accommodation?

The third session of the symposium was devoted to an analysis of the present strengths and weaknesses of the three major Bretton Woods institutions — the International Bank for Reconstruction and Development (the World Bank), the International Monetary Fund (IMF), and the General Agreement on Tarriffs and Trade (GATT). All three presentations which initiated the session shared a common viewpoint: only fundamental reforms of these three institutions will allow them to manage the developmental, financial, monetary, and trade problems which will confront the international economy in the decade of the 1980s.

In his analysis of the World Bank and its role in development finance, Mahbub ul Haq emphasized several problems whose resolutions would require major reforms in the rules and norms of that particular institution: (1) the rapidly growing investment needs of the developing countries (i.e., the projected need for $80 billion per year in energy investments in the less developed countries in the coming decade alone); (2) the apparent slowdown in Northern contributions to the "soft loan" window (the International Development Agency) of the World Bank at a time when, because of energy and food shortages alone, the ratio of needs to resources is growing rapidly in the world's poorest countries, which benefit the most directly from such "soft loan" sources; and (3) the prospects that the World Bank will turn away from both its analytical and lending efforts in meeting "basic human needs" at just the moment when it is beginning to master the difficulties of designing and implementing effective strategies aimed at alleviating extreme poverty conditions among the world's poorest peoples.

Underscoring his view that only fundamental reforms will produce acceptable results in the area of development finance in the 1980s, Haq concluded that only a World Bank whose expanding resource base was *internationalized* (i.e., *not dependent* upon contributions from nation-states) could meet the challenges presently facing the developing countries, particularly those with the lowest per capita incomes.

The second presentation, focusing on the international monetary system and the role of the International Monetary Fund, contained much the same sense of need for significant reform. In remarks rich in historical analogy, Paul Streeten argued that, compared to the relatively tranquil periods in international monetary relations characterizing the Pax Brittanica of the nineteenth century and the Pax Americana of the 1950s and 1960s, the present period is one characterized by "monetary schizophrenia." Relating this condition to the "pluralism" (or absence of hegemony) characterizing dominant features of the international economy since the 1970s, Streeten argued forcefully that the present set of international financial and monetary institutions are incapable of managing the problems those institutions were created to control. The difficulties inherent in present international arrangements led Streeten to offer seven very specific proposals for consideration, ranging from the incremental (an improved mechanism to "recycle" OPEC oil surpluses) to the fundamental (the need for a new international currency). In breadth as well as depth the Streeten proposals merit far more attention than time allowed in the session.

Finally, Gerald Helleiner echoed the call for fundamental reform, this time in relation to the world's trading regime and its major institution, the GATT. Despite his concern for some radically new rules and institutional changes, Helleiner's deep concern for what he viewed as the rapidly disintegrating rules and norms of the present system of liberal trade under the pressures of the "new protectionism" spreading in Northern countries led him to urge that Northern markets be kept as open as they presently are to the exports of developing countries. Critiques of Helleiner's views on the imperfections of the present trading system and his responses to those critiques produced some of the most lively and enlightening moments of the symposium. In the Postscript, further attention will be devoted to those discussions and the issues they raise for the future of North-South relations.

Session IV. Can a "Mutual Interest" Agenda Evolve during the Global Negotiation?

The two presentations which opened the final session offered an interesting contrast. Nicolas Barletta, focusing on economic issues, was generally optimistic about the prospects for the evolution of a "mutual interest" agenda which would, in the coming years, produce far less rhetorical conflict and much more concrete North-South cooperation in meeting the international economic challenges reflected in both the NIEO proposals and the emerging "global issues" — such as food production, new sources of energy, and environmental problems — raised by Walt Rostow in the initial session of the symposium. In sum, Barletta felt that the shared global objectives, and particularly the objective of returning to rapid economic growth, would eventually overcome the problems posed by present international economic conditions.

Ed Hamilton, on the other hand, appeared to give the nod to pessimism. He, too, stressed the elements of a "mutual interest" agenda which could be identified at the present time. For example, he suggested that the importance of Southern markets for United States exports had now become so dominant that United States commitment to present levels of trade liberalization might well withstand the growing protectionist trends that were of considerable concern to many symposium participants.

Nevertheless, on balance Hamilton chose to stress the "historical discontinuities" likely to dominate international economic and political relations in the 1980s, discontinuities which he linked specifically to the United States political and economic scene. His argument can be summarized as follows: (1) for the United States, North-South policies are almost always peripheral to and distinctly derivative of more salient issues; (2) in the next half-decade the two issues which will set the "limits of the possible" on United States responses to the South are the condition of the domestic economy and the confrontation with the Soviet Union; and (3) given the prominence of these two issues in United States politics, the probabilities for any major breakthrough in the North-South stalemate are minimal. The few areas in which he felt progress might be made, e.g., in some sectoral agreements and in debt relief, are detailed in his presentation.

Several final points should be made with reference to the symposium transcript which follows. The program was structured to build toward the question posed by Session IV: "Can a 'Mutual Interest' Agenda Evolve During the Global Negotiation?" And, as the reader will note, each session did in good part build upon previous exchanges. For this very reason the transcript of the symposium has not been heavily edited to avoid apparent repetition of issues previously discussed and points previously made. Many of the most interesting and revealing aspects of the symposium are to be found in nuanced but significant changes in phrasing and position. To trace these changes is often to reveal the impact of a broad spectrum of views on individually held opinions concerning crucial aspects of the North-South conflict.

Repetition of points of view that reveal *no change* have been left uncut as well, for a very specific reason. Such consistency of opinion and logic in the face of criticism is very often a revealing indication of the strength with which specific opinions and general perspectives are held. To have deleted this type of repetition would have limited the usefulness of this symposium as a document by which to understand the present North-South conflict as well as a lens through which to observe that conflict in action. For it is in the often fascinating participant interchanges which follow the introductory presentations that one observes the impact — or lack of it — of one viewpoint upon another.

Finally, it is worth underscoring the fact that the symposium contains not only an intriguing catalogue of conflicting opinions and perspectives, all of which help to reveal why the North-South stalemate continues to exist despite close to a decade of "dialogue" and "negotiations" to dissolve that impasse. The symposium also contains a very broad and interesting range of highly specific, concrete proposals for overcoming that deadlock and, equally important, for overcoming many major challenges to the efficient and equitable functioning of the international economy in the 1980s. At one point or another in the symposium, broad and comprehensive proposals are spelled out in considerable detail by Haq, Helleiner, Streeten, and Rostow; all other participants contributed proposals as well, if in less sweeping form. It is to be hoped that this balance — between the "ideal" and the "real," between the "philosophical" and the "pragmatic," between the critique of the present state of international economic and political affairs and concrete proposals to improve that state — presents a range of ideas, information, and perspectives that will challenge the reader as much as it did the symposium participants themselves.

Session I
The Setting

Elspeth Rostow: On behalf of the University of Texas, the Lyndon B. Johnson School of Public Affairs, and the Distinguished Tom Slick Professorship of World Peace, I am happy to welcome you to this symposium. This year, with an October meeting scheduled in Cancún, Mexico, on the North-South dialogue, we are particularly fortunate in having the Slick conference focus on this topic. We are even more fortunate in having as our Distinguished Slick Professor this year Roger Hansen, who comes to us from The Johns Hopkins School of Advanced International Studies in Washington, D.C.

The reason the University is happy to have the Slick Endowment is obvious: for any university, a serious discussion of the problems of peace should be an essential feature of its curriculum. Because of the scale of the problem, our emphasis shifts each year from one aspect of peace to another. It is therefore wholly appropriate that in the year 1981 we should talk about global negotiations.

I am pleased that you are here to participate in this discussion and now turn the proceedings over to Roger Hansen, the 1980–81 Distinguished Tom Slick Professor of World Peace.

Roger Hansen: I would like to thank the participants seated at the table today for their willingness to take two to three days from very hectic and heavy schedules to be with us here in Austin to analyze the issues to be raised in this symposium. I cannot imagine any other group of similar size that would represent more understanding of the "North-South problem" or commitment to finding constructive avenues out of the present diplomatic stalemate between the world's developed and developing countries than the group gathered here today. Again, therefore, my deepest thanks to you all.

Second, a brief word about the symposium program. The symposium will be held in four sessions. The first session will focus on what I have simply called the setting of the present diplomatic stage in North-South relations and, more particularly, the problems and opportunities presented by slowly altering perceptions, bargaining positions, and the goals of these two heterogeneous groups of states that frequently group themselves into two diplomatic blocs, often labeled "North" and "South."

In the second session, we will examine the issue of energy and its relationship to the North-South conflict.

The third session will examine the strengths and weaknesses of what we know as the Bretton Woods system — that is, the major institutions, norms, and governing rules of the post-World War II international economic system, particularly as they relate to the fields of finance, trade, and resource transfer to the developing countries. A crucial issue for the symposium here is whether the North and the South can overcome disagreements on the need for reforms of the system as it presently operates and can negotiate for an appropriate and mutually acceptable set of new or altered arrangements.

Finally, in the fourth session, we will try to pull together all these pieces of an extremely complex conflict — a political, economic, and psychological conflict, to highlight but three of its aspects — in an effort to see how sensible statesmanship might approach and deal with the set of problems encompassed by the present North-South impasse.

Each session will open with brief presentations from two or three speakers. Then the panel of participants will engage in a round-table discussion, as free and informal as possible, to analyze issues that have been raised in the presentations or that the panelists feel should have been raised and that are related to the topic of that particular session.

It now gives me great pleasure to open our first session by delivering the symposium into the most able hands of Ambassador Donald Mills.

Donald Mills: It has been almost seven years since a very important event took place in the world. In April of 1974, the Sixth Special Session of the United Nations General Assembly was held, at which the developing countries launched the formal negotiations and discussions on the issue of the New International Economic Order. Over the last seven years, there has been a great deal of discussion and negotiation in the United Nations and outside of it. Certainly, many of us are disappointed in the lack of concrete results of these discussions and negotiations. A lot has happened, such as changes in perception and the like, but in the face of the critical nature of the issue and the very serious state of the global economic system, we could have done far better than we have in the past seven years.

There are many people who believe that this is an issue invented by the developing countries, a set of demands put forward out of distress and frustration. On the contrary, it is an issue that would have existed and will continue to exist regardless of whether the developing countries had said a word about it. In my view, it is a historical imperative; it's a question of creating a balanced relationship between those people from one part of the world who, for whatever reason, have acquired a relatively comfortable level of living and those people, representing the vast majority of the world, who are in a state of relative poverty.

Of course, this is an oversimplification. But regardless of the causes of this great disparity, it is not possible to conceive of a sane world in the future in which we have failed to bring about a balance in the economic, political, and cultural terms. So whether developing countries have posed the issue in a sensible way is only one part of the problem. Responsibility for the lack of success in negotiations should be accepted by all — developing countries as well as the countries in the Western industrialized world and the socialist countries.

The foreword to this conference speaks of trying to reach some sort of consensus in these discussions. It also makes the very valuable point that, where there is disagreement, we should examine the reasons for the disagreement. In fact, this is one of the problems in the dialogue concerning the international system. We have not really forced each other to face issues, to face the possibilities of agreement, and to face the realities of disagreement. We have found all sorts of ways of avoiding that "locked room" situation where we all have to stand by what we say, changing our views if we are convinced otherwise.

Thus, much of the problem comes from the fact that Northerners disagree with us without listening. But please listen to us. When you have listened, you might very well disagree deeply with some of the proposals we have put forward, but that's the only basis on which there can be a real dialogue on the hope of agreement. This conference in a sense reflects that situation, and I would hope that we would face both disagreement and agreement in a constructive spirit and conclude with something which makes us all feel that it has been a very worthwhile exercise.

The term "global negotiations" refers to the process which the developing countries are trying to institute in the United Nations. But we have been involved, in fact, in a set of global negotiations for some time. We are now trying to improve procedures and raise the level of the institutional devices for dealing with the dialogue. There have been other attempts at dialogue, and the danger is that whenever we propose a fresh impetus or a fresh start after a previous failure, we come back to conflicts over two problems: the procedural devices to be followed and the agenda to be discussed.

One of the distressing things about having been in the system for as long as I have is that after a while you tend to say, "But I've heard that song before." We tend to fall back into the pattern of discussing, again and again, what mechanism and what agenda shall be used, and it's been repeated at least three times over.

The Group of 77 is the caucus of developing countries in the UN. It now includes over 120 developing countries that come together, whether in New York or in Geneva or in Rome — indeed, wherever there's a UN installation — to discuss economic matters of concern to the developing countries.

The Nonaligned Movement is a different institution. The Nonaligned Movement was not created in the UN. It was created by a group of Third World countries that came together in the late 1950s and early 1960s and formed a political association of countries with certain criteria for membership. As a result of the Cold War, which seemed to polarize the world around the conflict between the United States and the USSR, these countries decided that they, as developing countries, should have a separate voice and influence in world affairs. This political association has grown now to nearly one hundred members.

I don't know if I should attempt to give any other introductory definitions, but I suspect that the term "Third World" will be used often in this symposium. I think everyone will have some idea of what we mean in general terms when we use that phrase; most simply, it refers to the developing countries that have come together as members of the Group of 77.

I now have the great pleasure of introducing Mahbub ul Haq.

The Political and Economic Origins of the Global Negotiation of 1981

Mahbub ul Haq: Donald Mills has made my task easier by giving some perspective on the North-South dialogue, and I want to focus on how we can best move forward with the dialogue.

Before I offer seven concrete proposals on how we can move forward and what the agenda of such a dialogue would be, I simply must address a critical issue that Professor Rostow raised informally last evening. He has inspired me to rethink some of the issues on the New International Economic Order (NIEO) by raising the question of whether the current demand of the Third World for such an order has any intellectual, political, or moral basis. I believe that it does and that we should discuss this first. It will be very difficult later to carry on a discussion if some of the participants feel that the issue simply does not exist.

The *intellectual basis* of the Third World demands lies in the contention that international markets are neither free nor efficient today, but rather distorted. Through *de facto* and *de jure* organized intervention by rich nations, by either the governments or multinationals, the free workings of the market have been seriously distorted. The "free market" is a very convenient concept perpetuated by some discredited Western intellectuals in the waning days of that concept. They invoke Adam Smith without ever having read him and thereby do injustice both to themselves and to Adam Smith.

In fact, Adam Smith is the new guru of this intellectual school of the Third World. This school of thought suggests that there are three gross distortions in the present market structure.

The first is the restriction of the movement of people across national borders. This is one of the worst forms of denial of economic opportunities. It should be recognized especially by nations that were built very largely on migration.

Whenever people have escaped from our international "reservations," which is what the Third World is today, they have made a fairly good living on the basis of their own hard labor. In fact, when some of the economic frontier opened up in the Gulf states recently, just the earnings that people have sent home now amount, according to the latest World Bank study, to around $30 billion a year.

A second distortion is caused by restrictions on movement of goods. According to a recent World Bank study, $40 to $50 billion in exports each year in manufactured goods alone have been denied the developing countries due to Northern protectionism, and one can document a similar situation in the agriculture sector.

The third distortion involves restrictions on capital movements — the free movement of capital across nations.

According to this Southern school of thought, the rich nations are simply afraid to compete in truly "free markets." The aging nations are afraid of the young nations and their competition. So it is now the South that invokes "free market forces" as the rationale for asking for changes in the present international economic system.

Since present market structures, in this view, are dominated by very organized financial and multinational corporate power, this situation can be corrected only by organizing countervailing power in each field.

This school of thought has dressed up its arguments under various unsatisfactory labels: sometimes "deteriorating terms of trade," which is not the issue, and sometimes "relations between the center and the periphery," which has a little more ring of truth, but is still not the issue. The issue is the major market distortions noted above and the need to correct them. Galbraith understands this issue; Milton Friedman won't. I don't think we'll ever settle the argument.

Next comes the *political basis* of the New International Economic Order. The political basis lies in a struggle to share economic power and decisionmaking. It is a very logical sequel to the political liberation of most developing countries since World War II.

This school of thought demands, among other things, control over exploitation of a country's own natural resources and participation in international decisionmaking. Proponents of this view would take the very lesson of democracy back to its Western originators and preachers and argue that a majority of mankind is denied a significant voice in world affairs, whether in the World Bank, or in the International Monetary Fund (IMF), or in those quiet councils of the Group of Seven, constituting the so-called Western Summit States, in which major decisions are taken, without consultation with the Third World, about the world's major economic problems.

Those belonging to this school of thought will also say to hell with efficiency. This is not the goal. If, in getting greater control over our own economic destiny, efficiency suffers, so be it.

Political liberation was not fought for under the slogan of efficiency. At least there's an argument today over whether the new rulers in the Third World are running their affairs any better than the colonial powers did, but how much difference does that really make?

The North judges many proposals on the basis of economic efficiency; the South, on the basis of the control that can be acquired over decisionmaking. And as the objectives of these groups of countries are different, they naturally come to different conclusions about the same proposals.

Of course, one of the premises of this school is also that there is a Third World, which is a shaky premise. The countries of the Third World have a shared grievance and want to gain a place in the international councils, but they have not yet clearly defined their common interests.

One logical extension of this situation is greater South-South cooperation. Yet the Third World is not very well organized as a "trade union." It still has little self-confidence and its unity is too fragile, so these countries do not wish even to acknowledge the differences that exist among them politically, economically, and culturally. Nor do they wish to differentiate and "de-link" various elements in North-South negotiations and try to work on a functional, pragmatic basis rather than on an ideological basis. But all this will change in time. Political confidence and organization will develop. So in sum, the political basis for the NIEO demands lies in the South's perceived right to a greater sharing of power now held for the most part by Northern states.

And third, the *moral basis*. The moral dimension invokes the growing interdependence of nations and, as an integral part of that interdependency, a commitment to poor nations within the international community. It is the same principle that has been accepted within states — that a safety net has to be spread out for the poorest. The moral argument at the international level is the same — that there should be an international floor under absolute poverty.

Of course, these three schools of thought are not quite consistent. And different agendas will emerge depending on how the emphasis among them is placed.

The first school of thought, which is concerned with global economic efficiency, will stress competition, freer access to markets, and the use of countervailing power if it ensures competition on more equitable terms.

The second school of thought will stress political control, codes of conduct, new sharing of power within the Bretton Woods institutions, and the organizing of developing-country "trade union" power.

And the third school of thought will stand more for equity and primary concern for the poorest nations, and will invoke the "milk of human kindness," which appears to be fast drying up.

Naturally, these different Southern goals and emphases have led to a good deal of confusion in the agenda of the North-South dialogue and also in the tactics that have been followed by both parties to the dialogue. But one must not jump from this to a caricature of what the New International Economic Order is, a mistake I find so often in the literature. A common notion is that this set of Southern demands originated only in 1974 after OPEC's success. In fact, the struggle for a balanced political and economic relationship is a very old one, as Chairman Mills was suggesting.

OPEC merely demonstrated that in one particular area in which the West was very dependent, the existing rules of the game could be changed by some enhanced degrees of Southern-state organization.

So OPEC gave impetus to a movement that was already there. It's a caricature to say that this movement is based on the desire for a list of short-term concessions, a little more assistance, a little more debt relief, and so on. The NIEO stands for fundamental reforms in the present sharing of power in the world and the present international market structures.

It's also a caricature to say that the NIEO is being sought as a soft option by developing countries, a substitute for internal reforms. There may have been a tendency for that, and many of us in the South have criticized the lack of domestic reforms ourselves. But regardless of what is done within the South, the issue still remains: What kind of world order are we going to organize in the future decades?

So much for the polemics of the present North-South dialogue. Let me now turn to some quite concrete suggestions concerning what we can do to break out of the present acrimonious stalemate.

At the present time in global negotiations, the problem is both of tactics and of agenda. The South wants to link up every issue in an international negotiation, so that everything would change simultaneously. It is unwieldy, unmanageable, and unrealistic.

The North wants to negotiate case by case, function by function, and very often country by country. This approach is simply not acceptable to the South because these countries fear they cannot invoke any bargaining power in such fragmented negotiations. They feel very vulnerable on a country-by-country, case-by-case, sector-by-sector basis.

However, there is a vast area in between these two extremes that needs to be explored. This can be done by having concrete proposals that balance the interests of both sides and that have a good chance of political acceptability in the coming decade or at least in the next few years.

It's in this spirit that I will suggest certain proposals by way of illustration. These are proposals that will advance a number of the arguments the South has been making and yet will also protect the interests of the North and the needs of the world economy.

First, there is a need today for tremendous investment in energy resources. Everybody gains from that. In developing countries, the need is about $80 billion a year (in current dollars) over the decade of the 1980s.

Second, there is a mad search today for alternatives to oil, and everybody is experimenting with solar energy, biomass, and various other forms. There is certainly room for global cooperation here and for a consultative group on energy research in which various nations could jointly finance some of the research on alternative sources of energy.

Third, if we are to establish an international floor under absolute poverty over the next twenty years, we need to reorient foreign assistance to serve that purpose.

The much-discussed goal of a floor under "absolute poverty" obviously contains significant potential for a global compact between developing and developed countries on the financing and operation of an international poverty fund. To attend to the minimum basic needs of the poorest peoples of the world would require something like $25 or $30 billion of resource transfers plus a major effort by the developing countries themselves. That comes to 0.4 percent of the gross national product (GNP) of the rich nations. I am not very enamored of the target 0.7 percent or any of these targets established in UN debates and elsewhere because when we pursue them we find that they are, for most countries, merely rhetorical targets.

Even if the most generally accepted target of 0.7 percent of donor-country GNP were achieved today, it would mean little. For example, only one-third of Northern concessional assistance — so-called "official development assistance" (ODA) — actually goes to the poorest nations. The remaining two-thirds is being allocated according to political/security considerations. I would much rather have an international institution looking after the minimum commitment that humanity has to make to the poorest nations, so that this assistance is targeted toward an agreed-upon objective and efficiently spent, rather than maldistributed as at present.

Fourth, there are various ways of recycling funds today in an automatic fashion without putting a burden on the budgets of the developed countries. Changing the gearing ratio in the World Bank, which at present lends only one dollar for every one dollar of capital, is one way of doing it. The IMF's borrowing from the market against its gold holdings is another way of doing it. There are many ways of recycling market surpluses through active international intermediations.

Fifth, food security for nations and people is an urgent global priority. Can one resurrect any of the various 1970s schemes for food security? It will take $30 to $50 billion a year of investment in developing countries to increase Southern food production to levels that can insure food security over the long run. It will also take international programs of adequate food stocks and food aid.

Sixth, there are various ways to introduce *automaticity* into resource mobilization for development. Resource transfers must be made on a performance basis, but mobilization of the resources can become, to a limited extent at least, immune from daily political pressures. I have various favorite proposals to achieve this goal. Many people will think they are unrealistic, but I'm willing to argue that we shall witness some form of international taxation before the end of this decade. For example, I think it's possible to have an oil tax today and I believe there is significant progress toward an SDR link (that is, so-called "special drawing rights"). The forward movement on all these possibilities is far greater than many people appreciate.

And finally, South-South cooperation is going to be a key to the next phase of the dialogue. There are many things the South can do itself by organizing, particularly in the field of international services. A lot of the prosperity of the Western nations in recent decades has been based on development of services, and many of them are also *international* services of banking, shipping, distribution, and handling.

There is nothing so mysterious about this process that it cannot be organized by the Third World. If the Third World in the 1950s had only passed resolutions on industrialization instead of acting unilaterally and in groups to achieve industrialization, they would still be sitting in the United Nations passing meaningless resolutions.

Instead they developed their industries under protection and by tax concessions, and they came to a stage where they could compete in the international market and where, by now, developed countries are the ones asking for protection.

Similarly, today, they can develop international services. There is a major argument for organizing the Third World's own multinationals in many of these fields from shipping to distribution and giving them favorable treatment that would enable them to break into a field that has so far been exclusively Western dominated.

I will conclude with one word on the prospects of the North-South summit meeting that is to be held in Cancún in late October of 1981. This is a major development and should be very carefully prepared.

The summit cannot reach decisions on the specific kinds of proposals I have just been discussing. It would be a mistake to seek such decisions. Those should come within global negotiations, and special forums should be set up for each one of these proposals.

What the summit *can* focus on is what should be negotiated and where. It can examine whether today we are going to have global deflation as a result of deflationary policies in some of the Organization for Economic Cooperation and Development (OECD) countries; whether we are going to have growing degrees of bilateralism rather than strengthened multilateralism; whether this is the end of Bretton Woods and whether we should seek a different institutional structure for the future; and whether issues like international monetary reform, energy, and others that are now blocked in global negotiations are legitimate issues for serious international analysis, negotiation, and attempted cooperation in the early 1980s.

In other words, at the Cancún summit, we should try to set the *political agenda* and *forums of discussion* rather than try to make specific substantive decisions. The North-South summit should itself become a process: it should meet each year to monitor progress in North-South relations. If the summit accomplishes this, we may be able to make some progress in the months ahead.

Chairman Mills: Thank you, Mahbub, for laying the foundations of the discussion so thoroughly.

And now, Joan Spero.

The "Global Negotiation": Agenda, Progress, and Problems

Joan Spero: My task today is to talk to you in some detail about the global negotiations themselves — the agenda, the progress, and the problems.

I recently ended one year at the bargaining table of global negotiations at the UN, where along with Ambassador Donald McHenry I represented the United States in the negotiations during the Eleventh Special Session in August and September of 1980; and the continuing negotiations in a group called the Friends of the President during the Thirty-fifth General Assembly.

All of these groups were charged with reaching an agreement on a format for the Global Negotiation. I would like to describe briefly for you the nature of our charge from the General Assembly, and I would also like to try to explain to you, at least in part, why we were unable to fulfill that charge.

It was the Thirty-fourth General Assembly which passed a resolution calling for the launching of global negotiations. The resolution called for simultaneous, coherent, and integrated negotiations in five fields: trade, raw materials, development, energy, and money and finance. It charged the Committee of the Whole, known affectionately as the "COW" at the UN, with reaching agreement on the following: first, procedures — that is, how we would organize the negotiations; second, agenda — what specific items in these five fields would be addressed; and, finally, the time frame — how long the negotiations would last.

The discussion in the Committee of the Whole began in January of 1980 and was to be completed by the opening of the Eleventh Special Session in August of 1980. The two sides — and for reasons of simplicity, I'll talk about two sides, although in fact there are more — brought very different conceptions of global negotiations to the bargaining table. Let me describe those two conceptions briefly, admittedly in simplistic terms and in their most extreme form.

The Group of 77 offered a paper on procedures, agenda, and time frame for global negotiations. The proposal of the Group of 77 on procedures can best be characterized by the concept of centralization. As Mahbub ul Haq noted, the group wanted one central body or conference with the authority to negotiate binding and detailed agreements in all the issue areas. It wanted to use trade-offs among the issues in the bargaining process. It wanted universal participation, and, finally, it wanted to avoid as much as possible the existing specialized forums, especially the International Monetary Fund and the General Agreement on Tariffs and Trade (GATT), which it proposed to offer a consultative status.

The agenda proposed by the Group of 77 can best be described as the New International Economic Order, a long and detailed list of items in the five fields. Included in the agenda were important structural reforms, such as a change in the decisionmaking procedures in the International Monetary Fund. Also included in the agenda were a variety of items of concern to the different groups within the Group of 77, ranging from a commitment to the transportation and communications decade in Africa to the guarantee of financial assets of the capital surplus oil-exporting countries.

Finally, the time frame proposed by the Group of 77 was from January to September of 1980, or nine months, which came to be the time frame that everyone accepted while recognizing that it was unrealistic.

The developed countries offered a totally different conception of global negotiations. I should note that the developed countries never presented a comprehensive proposal as did the Group of 77. Rather, different countries offered various position papers emphasizing different aspects of the proposed global negotiations.

The reason for the absence of a common developed-country proposal — this is important to note and a point to which I will return — is that the developed countries in New York do not negotiate as a bloc as does the Group of 77. There is something called Group B in Geneva and in Rome, which is the caucus of developed countries (essentially countries of the OECD), but Group B simply does not exist in New York.

The developed countries, however, did have a general conception, if not a formal paper, of global negotiations. On the matter of procedures, in contrast to the Group of 77, they stressed decentralization, not centralization. They wanted to emphasize the role of the specialized forums, where those forums had legal competence and expertise. For example, monetary reform would be negotiated in the IMF, and commodity agreements in United Nations Conference on Trade and Development. They also felt that it would be unrealistic and unmanageable to try to link all of the issues, as proposed by the Group of 77.

As far as the agenda was concerned, in contrast to the Group of 77 proposals for the New International Economic Order, the developed countries took more of a problem-solving approach. They wanted to focus on priority items that in their view reflected serious problems and on which it would be possible to achieve concrete results within the time frame of nine months or a little bit longer.

For example, the United States offered what we called an "early action program" (and I must say it sounds a little bit like the program that Mahbub has just been describing). We suggested the negotiations begin by talking about energy production in developing countries, recycling, protectionism, world food security, and world food production. This general approach was viewed as insufficiently reformist and the specific agenda list as too short by the Group of 77.

We were able to move a long way from these two initial extremes. We made considerable progress in resolving our differences, but in the end, at the Committee of the Whole, the Special Session, and the General Assembly, we were unable to bridge the remaining gaps. As a result, there is to this date no agreement on the Global Negotiations. Let me describe how far we came before I suggest the reasons why I think we didn't get there.

On the issue of procedures, we developed the concept of a phased process that would combine both centralization and decentralization. The negotiations would begin with a conference, as proposed by the Group of 77, which would, by consensus, develop guidelines and objectives relating to the various items on the agenda. These guidelines and objectives would then be given to the specialized forums or to *ad hoc* groups that would be set up to carry out the detailed negotiations. This latter procedure responded to the Northern preference for a decentralized approach.

After these groups carried out their negotiations in a decentralized fashion, the results of each separate negotiation would be returned to the conference to be assembled in a final, so-called "package agreement." Obviously, this final step represented another concession to the Group of 77 concept of linking issues and making trade-offs at the end of the day.

This, then, was our general agreement, but we still had important problems with the details of that agreement. The major problem, in my view, concerned the balance between the role of the conference and the role of the specialized forums. How could we satisfy the desire of the developed countries to preserve the autonomy and the utility of the specialized forums while also satisfying the desire of the Group of 77 to bypass the specialized forums in order to achieve their desired goals?

Another important procedural barrier related to energy. Energy, it turns out, is the one issue with no specialized forum. It is the one issue to which there was no logical venue for decentralized analysis and negotiation. Furthermore, the oil-producing and oil-exporting countries feared greatly that global negotiations might, through the process of decentralization, lead to the creation of an energy institution that they have very much opposed ever since the issue first arose in 1974.

We were able to reach some agreement on the agenda by making it more neutral. This was accomplished by emphasizing the goals of the New International Economic Order in a preamble while developing the actual working agenda in more neutral terms. Important differences remained, however, especially in the areas of energy and money and finance.

Finally, on the issue of the time frame, as noted earlier, everyone agreed that the entire process would last nine months; ironically, everyone also came to agree that this time frame was unrealistic.

The main question I want to address is why we were unable to bridge these gaps and why there has been a continuing stalemate regarding the launching of the Global Negotiation. There are many levels of explanation for the failure of negotiations. They range from the concept of disparity of power among the participants in the negotiation to the fact that an election took place in the United States during the process of the negotiations. I would like to offer a perspective from the negotiator's standpoint, a microperspective of the problems as I saw them in the process of trying to negotiate them at the United Nations.

My main point is that the *process itself* is a barrier to agreement. Let me explain what I mean, and let me say quite frankly that I give you a slight caricature of the process in the hope that I will provoke some response from the other participants.

Let me look at this process at three levels: first, the individuals who participated in the process; second, the groups that organized the process; and, finally, the bargaining structure of the United Nations itself.

The participants in the negotiations, the key actors in this process from the North and the South, are very different. The reason they are different and bring very contrasting perspectives to a negotiation is an outgrowth, to a great extent, of the way governments are organized to make and to execute North-South policy.

In the simplest of terms, the South is represented by diplomats and representatives of foreign ministries, while the North is represented by economists who come not only from foreign ministries but also from treasuries and finance ministries. One very significant result is totally different views of reality and totally different conceptions of what the negotiations are all about.

First, the Group of 77. Again, I apologize for making rather sweeping generalizations. The activists behind the Global Negotiation are diplomats from the foreign ministries. In fact, very often the activists are diplomats based in New York, because the foreign ministries in the capitals, for a variety of reasons, are not intimately involved in the negotiations.

The diplomats from the Group of 77 are not trained economists. They adopt the economic analysis that argues that the system is fundamentally distorted from a development viewpoint and is inequitable, although they are not always able to argue the case effectively about the distortions in the market. Furthermore, the representatives of the Group of 77 distrust the International Monetary Fund, the General Agreement on Tariffs and Trade, and, dare I say, the World Bank. Although many of them, because they are not specialists in this area, do not fully understand the operation of these institutions, they are persuaded of the political argument that it is necessary to change in effect a fundamental system.

And here I want to add a brief footnote. That is, the message heard from the diplomats in the United Nations is not always and very often not at all the message that is heard in the capitals by the representatives of the developed countries. The talk of the New International Economic Order, the talk of a multilateral approach to reform, the discussion of fundamental changes in the system — this is not the message that American officials hear in the capitals of the developing countries.

Finally, the representatives of the Group of 77 see the problem in achieving the New International Economic Order as one of inadequate political will in developed countries. Those who argue against the proposals in various dimensions of the New International Economic Order are viewed as self-serving. The Southern viewpoint is that with the proper "political will," all would be relatively easy.

The developed-country activists, on the other hand, are by and large economists. They are most often stationed in their national capitals rather than the UN, and UN diplomats from the developed countries are kept on very tight leashes. The key actors in the developed countries therefore generally see North-South issues in traditional terms of economic efficiency; that is, in terms of preservation of the market and of the existing institutions that presumably support that market. They reject the economic analysis adopted by the Group of 77. They view the system itself and its institutions as having a fundamentally positive value, not only for the developed countries but also for the developing countries. While they recognize the existence of imperfections in the international economic system, they believe quite honestly and quite seriously that all stand to lose by restructuring along the lines of the New International Economic Order.

Furthermore, the North-South dialogue is a relatively low-priority item for these foreign policy economists. They have to pay attention to a wide range of issues, from the problem of the dollar to Japanese auto imports. For them, the North-South dialogue seems "unrealistic." It attracts little of their attention and, more importantly, little of their creative thinking and creative effort. This low-priority status is reinforced by what they hear from the capitals of the less developed countries (LDCs), and by what they hear from LDC representatives in IMF and the World Bank, where there is little talk about global negotiations.

In sum, and again in caricature, they have come to view the UN as a dangerous place where diplomats have different and, from their perspective, misguided opinions concerning economics and the system which, if acted upon, can do serious damage to the world economy. Therefore, to the extent that they want to be responsive to the South, they want to be responsive to the South outside of the UN context. Thus, their preference for decentralized, issue-specific negotiations within the IMF, the World Bank, and the GATT.

The result in the negotiations, then, is a fundamental distrust on both sides about the intentions of the other side. What did they really mean? What are they really trying to put over on us? There is a lack of communication, and there is a polarization and a rigidity in these views.

The group structures of the United Nations are also a problem in the process. The groups are important at the UN in facilitating negotiations; quite simply, it would be impossible to conduct a negotiation among 154 countries. But the groups also pose important problems and impediments in the negotiations.

The most important group is the Group of 77. It is well organized to express the economic concerns and the programs of the less developed countries, but it is poorly organized to bargain and to negotiate. Unity is the key asset of the Group of 77, and that unity is achieved by including demands of a great diversity of groups. Furthermore, the Group of 77 is a democracy in that each of the various groups with diverse viewpoints has a veto power in the system. Thus, it is difficult, if not impossible, for the group to alter its proposals without losing important support within the group and without threatening to fragment the unity of the group. But altering proposals and changing positions is the essence of successful negotiation.

The leadership of the Group of 77 is faced with an important dilemma: Can it make deals at the bargaining table which it can eventually sell to the group at the end of the day? The answer, unfortunately, is often no.

Let me mention one other special group, OPEC. It is a crucial group within the Group of 77, or at least it was during the process of planning for the Global Negotiations. The discussion of energy is a key element of global negotiations, but OPEC had great troubles in contemplating an energy dialogue.

OPEC — and it is perhaps unfair to talk about OPEC as a unit, but let me do this because of limited time — did not want energy to be discussed in an isolated forum. Because OPEC feared the isolation and the heat of discussing energy alone, it wanted energy linked to a variety of other issues.

Furthermore, OPEC did not want to discuss recycling. Recycling was, as we started to call it, a four-letter word in planning for the Global Negotiation. We were not allowed to use the term. And OPEC also did not want to talk about energy supply or energy price.

It was perfectly willing to talk about energy demand. It thus played an important blocking position within the Group of 77. Indeed, the question was raised whether OPEC wanted global negotiations at all.

Although the developed countries have similar interests, they do not negotiate as a group in New York. The result of this situation was that their positions often differed. They often spent as much time haggling and negotiating among themselves as they did negotiating with the Group of 77. They often sent confusing and different signals to the Group of 77, which led to disorganization and confusion in the negotiations.

Finally, there is the third-level problem of the structure of the negotiations themselves. The regular organs of the United Nations are rather well structured for bargaining and trying to achieve compromise. The Security Council has a regularized system of consultation. It has a method and a structure that leads to or at least encourages compromise. The same is true in various committees of the United Nations General Assembly, where there are regularized consultative processes.

None of that institutional infrastructure existed for the earlier negotiations, and the fact that the negotiations were relatively unstructured posed, I believe, terrible problems.

One of the major problems was the question of who negotiates — that is, who comes to the bargaining table. It is impossible to negotiate with 154 countries, but negotiating with any fewer is often viewed by those unrepresented as illegitimate.

A related question became whether the chairman of the Group of 77 should come alone to the bargaining table. That is acceptable, but if he does not bring some of the members of the group, it is highly unlikely that at the end of the day he will be able to sell his achievements to the rest of the members of the group. If he brings a few members of the group, who should come? If you are not invited to the bargaining table, will you pose problems? Will you cause problems for the chairman if you are left out?

The European Economic Community had similar problems. The EEC negotiates as a unit, or did so on the earlier global negotiations. But, of course, all of the ten want to be present at the bargaining table. That's understandable because they are the key economic powers, and they want to be in the room. But if they all come, then all of the Group of 77 have a right to come and, once again, you're in a large room.

The United States, I should add, has no problem. We are always invited, for obvious reasons.

All of this puts a major burden on the chairman of the negotiations. The chairman becomes responsible for, in effect, organizing, managing, and orchestrating the negotiations. He is the arbiter. He is the prodder. He determines who participates. He often assumes the risk of producing compromise papers.

If you have a good chairman, you may have a good negotiation. But a good chairman is a very weak reed on which to build a negotiating structure, for if there is no good chairman, the negotiations can rapidly and easily flounder.

Therefore, I personally conclude that the *process* is an important obstacle to the negotiations. And even if we reach agreement at Cancún or at Ottawa or elsewhere, the problem of the process at the UN will remain and will be, in my view, an important obstacle to the Global Negotiation itself.

Chairman Mills: And now, Walt Rostow.

Beyond the Official Agenda: Some Crucial Issues

Walt Rostow: As I told some of my colleagues last night, I am in a rather awkward position because my view of the New International Economic Order negotiations that have taken place since 1974 is that they have been based on the wrong intellectual conception, the wrong agenda, the wrong negotiating forum, and the wrong cast of negotiators. I think that in different ways both of the speakers this morning have illuminated why that perspective is a defensible view of the dialogue since 1974.

I perhaps should start by explaining my broad perspective, which leads me to conclusions that are quite different from those of my hard-working and responsible colleagues who have been engaged in this process.

While they were working through this difficult and important process, I was fulfilling a youthful commitment to write a history of the world economy covering the past two centuries. As I worked forward to the present and looked to the future, it became clear to me that we had entered, at the end of 1972, a new phase in the history of the world economy. This phase, parallel to four previous phases, will be marked by relatively expensive raw materials.

The list of resources differs with each of these phases, but in this case it's clear enough what the key relative shortages are. They are energy and food. And if we achieve high rates of growth, other costs will also rise rapidly, most notably those for clean air and water and others relating to problems of global pollution. That's not a total list, but that's a fair way to begin.

What I found so discordant as I watched the unfolding of North-South negotiations was the fact that they seemed to be based on an obsession with certain broad issues which were, in some sense, relevant, but which were totally out of touch with the reality of the problems that were grinding down the whole process of growth in the developing world during these same years.

My commitment to the developing world goes back at least thirty years. And what has concerned me is that, while these high-level negotiations were going on, the rates of growth in the developing world were decelerating in increasing proportion as the foreign exchange of the developing countries was being diverted — at great cost to their development — to the purchase of oil and increasing amounts of food from abroad.

They were failing to solve the problems that would produce adequate flows of capital and develop their basic resource for their own use and for export to the world. Meanwhile, there were very grave problems of the degradation of the environment: soil and forest erosion as well as the inability to cope with the urban environment.

So I honestly believe that the devoted human beings who have been engaged in these negotiations have not been serving well, through no fault of their own, the men, women, and children of the developing world.

I am going to suggest four key areas that deserve priority attention. To start with, there is a need to build a new energy base in the developing world, including most of the OPEC countries, which are shielded only temporarily by the high oil prices from the reality that they, too, will have to build this base. There is also a need for an acceleration of agricultural production and programs of nutrition for the young, especially including the provision of cheap protein, which requires something like an average 4 percent rate of growth for agricultural production in the developing world.

A third key area concerns measures of aid to buy time. We've got to buy time for the small, poor, vulnerable countries faced with high-priced imports of oil, degradation of the soil, and excessive population pressure. These are, in my view of the global requirements, welfare problems that we've got to face.

Finally, there are certain large countries that are heavily indebted. Brazil is the most notable, and although it is making extraordinary efforts to solve its energy problem, it will have to be given time to do so by the international community. There are also others, so there are certain aid tasks to be faced.

If I were one of the Group of 77, I would raise hell with the North. I would say that you have no right, given your global responsibilities and power, to deal with energy as you are — in such a foolish manner. You've got the resources in coal and oil and gas and shale and potential nuclear power. Make yourselves much more independent of oil imports and leave more for us, because the developing world has much higher rates of growth in energy consumption. We can eventually curb these growth rates and in so doing put a ceiling on the oil price, which is cutting our throats in the developing world.

I have proposed, for example, that the United States should seek to achieve a net energy export position by 1990, so that we can, in part, gain control of our own domestic economy again and, in part, render our strategic position more flexible. But such an achievement would also give greatly needed breathing room to the developing world. While this policy goal is surely achievable, we're not doing it, and neither is Western Europe.

I have also proposed that the Group of 77 ought to be insisting that the North solve the problems of stagflation. It's ridiculous to go on running economies at low rates of growth and trying to cope with inflation the way we are. And I have no optimism whatsoever that we're going to get generous aid terms or be able to fight off protectionism very effectively if we have another decade of acute stagflation in the North.

The North's two fundamental responsibilities to the South, energy and the conquering of stagflation, belong on an honest agenda. You can add others. I recently participated in an exercise of the Organization of American States (OAS) designed to define an agenda for hemispheric cooperation in the 1980s. I asked my Latin American colleagues on the first day what they thought should be on our agenda. The agenda that emerged was fascinating. It included energy, food and nutrition, raw materials, the environment, and then some interesting topics that I have not mentioned: the definitive overcoming of the problem of illiteracy as well as the build-up of Latin American capacity in science and technology. This functionally specific resource agenda fell right out on the table, as it will with any sensible pragmatic people looking at the world today.

Now, I submit that there is no way you can handle this agenda on a global basis. The most this kind of global enterprise can achieve is the writing of an agreed resolution, leaving each government to interpret it in a way congenial to its interests. Now, if the resolution reflected an honest consensus that these were the tough issues, you might then put people to work. But what should you do if you actually got serious? What actions would follow?

What should be happening is this. The countries in the developing world should be developing ten-year energy programs that would meet their energy requirements with a maximum cut in energy imports, diverting their own resources to these programs and drawing capital and technological assistance from the outside world. This would enable them to develop their own resources, an energy base consistent with their high rates of growth. This means cooperation not on energy in general, but on coal, nuclear, oil, gas, solar, biomass, etc. In biomass the Brazilians are leading the way. There is a lot of potential for South-South cooperation, not only in finance and services, but also in technology.

You would thus develop a series of energy plans, and then you would figure out how to help finance them domestically and from abroad. In agriculture, the same approach would apply.

Now, you can't sit with 150 states in the United Nations and work out general energy and agricultural investment plans and specific projects that have to be carried out to fulfill them. Nor can you do it with ambassadors or even with economists. You've got to deal with the responsible officials in agriculture departments. You've got to involve the heads of all the relevant ministries in energy, agriculture, education, etc.

Furthermore, as I suggested earlier, energy is not one thing; it involves many potential sources. You've got to get the best people on nuclear power, the best people on biomass, the best people on tarsands, and the best people on hydropower. If you're serious, that's what this resource effort is going to look like.

The commodity agreements would be global and broad trade agreements would be global, but I think the best general solution would be regional in orientation. You could get the Organization of American States (OAS) to play a major role. Europe should lead the way in working with Africa, where the problems noted above are hideously urgent while all of this palaver goes on.

It's also time for the Pacific Basin to get organized. The Pacific Basin represents the most dynamic part of the world, and it's developing a network of interconnection at least as sophisticated as that of the Atlantic world.

The regional agendas would be functional. There would be plenty of chance for cross-functional exchanges. The World Bank, of course, would work with all regional institutions as well as the IMF, the Food and Agricultural Organization (FAO), and other global organizations. But you would get down to work regionally and start generating investment projects that would cope with these acute resource and related balance-of-payments problems. The only way they're going to be solved is by building up the resources in energy and food to free the balance of payments for the growth processes that must go forward.

The OPEC countries, the oil exporters, would play their part in their regions so far as new energy sources are concerned, but the issue of price should be ruled out. The only way we're going to get the OPEC price in some sort of order is to build alternative energy resources and to conserve in such a way that market forces bring down the real price of OPEC oil.

We published a responsible paper at this university.* No one paid much attention to it, but I believe it to be correct. Its author said that using existing technologies we can produce oil from coal at a $22 per barrel price equivalent (in 1977 prices), but at a 10 percent discounted cash flow rate of return. If we did it on a large scale and got on to a learning curve, we could bring that price down considerably. That's probably the biggest contribution the North could make to the South: to put a ceiling on the oil price and start bringing it down. But you're not going to do it by talking your way to agreement with OPEC or with other oil exporters. You're only going to do it the way these things happen in real life — by developing cheaper alternative sources on a large scale.

That's the way I think we have to go about resolving today's crucial problems. I never believe that the ideas I generate come down from a mountain in marble like the Ten Commandments. My test of whether I may have the correct solution in sight is whether others are arriving at it on their own. Embedded in that rather overly complex and multidimensional Brandt Commission report, which had a diffuse impact, were sections on oil, energy, food, and even regional cooperation with which I would agree. And I was much heartened when I saw my strong-minded colleagues in the Western Hemisphere emerge with an agenda very much like the one which I had deduced from my own studies. And I am much heartened that Mahbub ul Haq and Joan Spero referred to movement in those directions moments ago.

---

*W.C.J. van Rensburg, "Coal Gasification and Liquefaction," in *National Energy Policy Issues* (Austin: Council on Energy Resources, University of Texas at Austin, June 1979), pp. 227-268.

I hope that we get on with this pragmatic agenda. If we would look at the balance of payments of the developing countries and their inadequate energy and agricultural policies and see the human cost, the social cost, and the cost in political disruption of failing to face these problems together, we can make a pretty good case for getting on with this sort of agenda.

In the end, we will find that the classic objectives of the New International Economic Order will reemerge out of big investment programs focused on these sectors that are so vital in the next quarter-century.

Chairman Mills: Mahbub ul Haq spoke of the intellectual, political, and moral basis of the call for the New International Economic Order. I would mention that there are some in the North and South who would perhaps underline a bit the issues of self-interest and self-preservation. They would look at things like population figures and wonder how you could survive in a world in which these extremes existed. There is also the question of mutual benefit. There have been attempts to calculate how much benefit Northern interests would derive from some of the proposals that have been put forth.

Nevertheless, there are some Western countries that have agreed in broad terms, not in the details, with the proposals of the developing countries and the general idea of the need for a New International Economic Order. The Nordic countries and the Netherlands are among them. It's interesting to ask why they agree and why they have agreed from the beginning. There are all sorts of reasons, but this topic is another dimension to the whole question of the failure to have a more successful dialogue with the main body of Western countries.

Mahbub ul Haq has offered a series of proposals that are subjects for discussion. He has also mentioned the North-South summit and the things that summit should and should not attempt to do.

Joan Spero gave us a very incisive analysis of the roles and behavior of the various participants in the negotiating process. Although I would disagree with her on many specifics, she is right in saying that the process itself is an obstacle to reaching an agreement. I hope we'll be able to discuss that issue to some extent.

Instead of approaching the New Economic Order as a vast, nebulous, global undertaking, Walt Rostow invited us to approach it from the perspective of what he calls a resource-oriented, functional, pragmatic agenda that includes energy, food, and other vital issues.

Of course, he argues that South-South cooperation and regionalism are important elements. He also stresses that we should make sure we have the right cast of actors — that is, the people who are going to make the decisions and implement them rather than lower-level diplomats and economists from relevant ministries. That, again, I happen to agree with very strongly.

I don't agree necessarily with the proposition that we come back to the New International Economic Order at the end, that we might arrive at it after we have been through the process he describes. Many of us are trying to start with it. It's a very interesting concept.

Altogether, then, we have a very good basis for discussion, and I invite comment from our group here.

Panel Discussion

Paul Streeten: I want to take up the challenging question that Walt Rostow asked last night and that Mahbub ul Haq answered so well this morning relating to the three bases of the New International Economic Order: the intellectual basis, which contains the economic; the political basis; and the moral basis.

First, I would like to invite Walt Rostow, a distinguished historian, to add something about the historical basis for the NIEO, for I think it is useful for a full understanding to analyze the historical origins of this set of demands. My own inadequate analysis would identify three historical reasons for the NIEO demands of the developing countries.

The first one is the disappointment with political independence. It has not given many ex-colonies the kind of freedom to act and promote their objectives and aims. In other words, political independence did not produce the hoped-for economic independence.

The fact that many of the claimants to the NIEO are Latin American countries that have been politically independent for a long time does not, I think, refute the point but in a sense confirms it. Many Latin American countries do feel economically dominated. This is, after all, where dependency theory has come from.

A second reason results from the disappointment with the aid flows, with the transfer of resources on concessional terms, with the amounts, with the conditions, and with the kind of relationships that aid has given rise to.

If concessional aid flows had moved toward a genuine international income tax, toward a genuine redistribution of resources on an international scale and for the right purposes, I think the demands for the NIEO would have been somewhat less sweeping. They certainly would have been slanted in a different direction, not that of "concealed" flows through raising prices and debt relief and other measures of this type.

The third reason is the success of OPEC in transferring a substantial amount of resources by collective but, in a sense, unilateral action, that is, a group of developing countries getting together and acting in concert.

The notion that this kind of action can be applied to other types of commodities has been, on the whole, a disappointment. Many people would argue that oil is unique and that this transfer of resources, by stabilizing, as they call it, but in fact by raising, prices of commodities, cannot be accomplished in other fields.

It's useful to remember the historical basis, not so much as history for its own sake, but in order to understand where these demands come from and how people in the North can respond more constructively.

It is not enough for economists from the North to deal with the representatives of the South as if they were rather poor students who hadn't done their economics homework. I think many of us around this table would agree that the NIEO demands and priorities were not terribly well thought out. I don't know how many of us would have put so high on the NIEO agenda an integrated commodity program that essentially consisted of stabilizing and raising prices of commodities that are not necessarily exported by the poor countries or beneficial to the poor people. But instead of saying that the South is wrong time and again because of faulty economic analysis, it would have been more constructive to say, "Well, this may be the wrong way of setting about our shared aims, but let us think of some better ways of responding to these demands."

I don't think the North has done that. The North has been very, very negative and very obstructionist in regard to the NIEO demands. Perhaps Walt Rostow will tell us how he understands the origins of the NIEO and what he considers more constructive responses.

My second set of remarks concerns the moral basis of the NIEO. I don't know whether you like using the word "morality." It is not very popular in many circles, but I think it is important to explore this issue more thoroughly and to compare this aspect of the demands for the NIEO with the pragmatic, specialized, compartmentalized approach that Walt Rostow has suggested. But let's not call it "morality." Let's call it "international solidarity" or something like that.

Anyone wanting to analyze this aspect of the New International Economic Order has to answer three questions, none of which is very simple.

The first question, if we assume that we live in a community of human beings, is whether the rich in that community have any obligation toward the poor, and whether the poor have a just claim to the wealth of the rich. There are very respectable philosophers who would answer that question with a no.

Robert Nozick from Harvard would probably answer as follows: "Well, if I earn my money and my income selling goods in a legitimate way, I am entitled to hold onto my earnings. I have no obligation to do anything for the poor and the poor have no claim on me." I think this is essentially a mistaken view. The rich do have a moral obligation to other members of their community — to help them and to produce the kind of safety net for the poorest that Mahbub ul Haq has talked about, whether on utilitarian or Rawlsian grounds.

The second question is more difficult and more controversial than the first one: Does the present world of nation-states constitute a "community" in the relevant sense for this type of purpose?

Many people who would say yes to the first question might say no to the second one. They say that people living in England are too far away from people in China, and that people have to look after the poor in their own neighborhoods, their own cities, and their own countries first.

But I think most of us would agree that the world, in some sense, is a community in which the rich have obligations toward the poor, though possibly weaker ones.

The third and perhaps most difficult question is one which the debate about the NIEO has brought out very clearly if not always explicitly. Given the fact that the rich in the rich countries have a duty toward the poor in the poor countries, how do we deal with the fact that the obstacle in our way is the nation-state and its government, which often acts inequitably? Doesn't the fact that the world is constituted by sovereign nation-states interfere with the moral obligation of the rich people of the world to help the poor?

It has become a cliché to note that aid often consists of transferring resources from the poor in the rich countries to the rich in the poor countries. This proposition probably came up first in the aid debate in England, and it contains the kind of half-truth that is particularly galling to those of us who believe in the other half of the truth.

This issue raises a host of other issues about relations between

Davidson College Library

39

governments and relations between people. How can one ensure that the agreements between governments will ultimately improve the lives of the human beings and their families that we are concerned with? National sovereignty implies the right to use aid funds iniquitously.

This question has been somewhat neglected in some of the negotiations. It has also been neglected in the Brandt Commission report. The report has a lot to say about international as opposed to domestic issues; it was, of course, a commission established for that purpose. Nevertheless, it talks too much about intergovernmental relations and too little about human relations.

Let me conclude by saying that I cannot help but agree with the pragmatic items on the agenda that Walt Rostow has put forward: energy, food, raw materials, and population. Yet I'm worried that this kind of approach may fundamentally conflict with the moral or solidarity approach. If we solve the global problem of energy, for example, of course we will contribute in some way, somewhere, toward the relief of poverty.

But I suspect that the impact of the solution may be highly capricious and uneven in the distribution of benefits. It would differ radically from the approach of placing high on the agenda of international cooperation such issues as the elimination of poverty, of malnutrition, of hunger, of starvation, of disease, and of illiteracy.

So the kind of approach that begins with joint and shared objectives and then mobilizes support for solidarity to achieve those objectives is a different kind of approach from the one that emphasizes increasing the supply of food and energy, and mobilizing the political and financial support for these functionally specific sectoral tasks.

I think we could use both approaches. The pragmatic, sectoral, specialized approach could complement the "solidarity" approach. But I don't think either one is sufficient alone.

Chairman Mills: Thank you, Paul. Now, Professor Helleiner.

Gerald Helleiner: It strikes me that the sense of history that Professor Rostow insists on is even more important than he himself makes it. It may be that in his own presentation he understates its importance, because one of the things that pervades the kind of implicit dispute between Joan Spero and Mahbub ul Haq is the relative difference between emphasis on short-term crises and emphasis on longer-term historical changes.

In the lists of proposals we have heard, the one that Mahbub suggests and the one described as being the essence of United States proposals, the most striking overlaps are those that relate to universally agreed-on current problems, such as recycling, energy, food, and protectionism.

The disagreement relates to the longer run. Mahbub's proposals have to do, for example, with automaticity of the mobilization of resources for development. By the year 2000, I think, that proposal will be a reality.

He also proposed setting a floor under global poverty. I believe that by the year 2050, our grandchildren or their children will be amazed that we had no such thing at the global level when we already had something of that kind at the national level.

It's at this level that it's important to distinguish the short run from the long run. The Brandt Commission, for all its faults, attempted to do that by distinguishing emergency food, recycling, and energy programs to deal with the immediate poverty problems because there are now people in danger of severe suffering and of starvation in the short run. Yet, it's very important not to lose sight of the longer term while grappling with these enormously difficult short-term problems.

In dealing with the short run, it's also important to understand that the North will set about putting its own house in order, whatever the South does. The Western economic summits will seek to solve stagflation, but they don't need advice from the South on the question, and nothing that the South says on the subject will alter what they do. Under the present system, they will make their own decisions.

It is for the South to press continually for an agenda that includes those portions of the international discussion that *would not otherwise be placed there.* This is true of the resolution of short-term difficulties as well as the long-run ones. The long-run problems are those which are most in danger of not being placed on the agenda by pragmatic politicians constantly needing to worry about being elected four or five years hence.

My second point is that we can confuse New York with the world in this process of global negotiations. To a great degree — and I mean no offense to our ambassadors — the United Nations in New York is a theater where neither the North nor the South is represented by its highest officials, whether in economics or in other spheres, in the debates which take place.

When it comes to negotiating the details of an international tin

41

agreement or the terms of the code in the GATT that governs customs evaluation, the North has more than its share of the lawyers, and the South, through no fault of its own, is not as well equipped to staff the enormous number of meetings continually going on under UN auspices and as part of the Global Negotiation.

The Global Negotiation consists of far more than what's going on in New York, and happily it has a momentum of its own. Discussions will proceed within the existing forums — the IMF, the GATT, and everywhere else.

Chairman Mills: Thank you. Jorge Navarrete.

Jorge Navarrete: At this stage, I would simply like to add a word on the problem of the basic rationale for the New International Economic Order.

I think there is a combination of factors that constitutes the basis for the New International Economic Order. The economic, political, and the moral factors mentioned here today are mixing together to establish a base of common interests among all members of the international community as they seek to restructure international economic relations.

The experiences of the 1970s concerning the behavior of the world economy and the disappointing results of international economic negotiations are forming a new basis for renewed efforts. Perhaps now, at the beginning of the 1980s, it is clear that interdependence in world economy is stronger than at any time heretofore.

It now seems clearer that no segment of the world economy can really aspire to growth and to prosperity if these conditions are not shared by the other segments of the world economy. To note but one example, in the mid-1970s the import demand generated by some of the relatively more advanced developing countries was one of the main elements in the economic recuperation of the OECD countries. In my view, this is one of the first manifestations of a new form of interdependent operation in the world economy. If this happens increasingly in the future, there will be a strong, common interest in reforming and restructuring world economic relations by taking into account these new realities.

Chairman Mills: Thank you, Jorge. Sidney Weintraub.

Sidney Weintraub: Let me make a comment on process: the bargaining process is going on all the time at literally thousands of levels. It takes place in functional institutions, it takes place bilaterally, it takes place multilaterally, it takes place regionally, it takes place globally. It takes place globally in New York, in Geneva, and in other places, depending upon the issue.

Therefore, if we talk about North-South negotiations only under the rubric of "global negotiations" or "New Economic Order," we're really missing the essence of what goes on in the international arena.

Let me just confirm one point that Joan Spero made. During my years as a bureaucrat in the Economic Bureau of the State Department, which is essentially concerned with international trade and international money, we paid very little attention to what was going on in New York. It was not because we were professors scolding poor pupils. That really had nothing to do with it. What really mattered was our perceived self-interest — where United States trade took place and what the world monetary system was like — and we therefore devoted most of our time to those issues rather than to the issues in New York. I suspect that has not changed.

My second point is a fairly simple one. Part of the problem that many of us in the North face is the premise that there is a Third World. I think we have to recognize and deal with the vast difference between a Brazil and a Chad. What Latin America is interested in may not be the same thing that Africa is interested in. The assumption of unity of motive always bothers people from the developed countries.

My third point concerns the notion that there has been no progress in the world in the last two decades. I just don't believe that. The growth rates since World War II have been higher than at any other time in recorded history. Of course, this growth has not been universal. The very poor countries have not shared in it, but the middle-income countries, the new industrialized countries, really have very little cause for complaint about the way the global economic machinery has been operating. There may have been little progress in New York, but there has been progress in many other arenas, as growth records suggest.

Finally, I don't think the issue is really a lack of political will, as people sometimes put it. I once defined "political will" as whatever you think the other fellow should do but does not want to.

The issue really is substance, and the issues of substance determine the way nation-states operate. Unfortunately, what interests them is their own self-interests, and therefore the substantive issues really have to do with income, employment, and then those areas of commonality in which income and employment benefit all of us if we work together.

I think people would probably agree that we ought to work much more carefully in laying a floor under global poverty, figuring out a better way to recycle petrodollars, increasing energy production, and trying to increase food production. We agree in principle but not always on the details.

Where the countries of the developed world and the countries of the less developed world disagree is where the self-interests of the two diverge, and that divergence of self-interest explains most of the conflict over the New International Economic Order.

Chairman Mills: Thank you. Roger Hansen.

Roger Hansen: I must disagree with my good friend Sid Weintraub, at least on a matter of emphasis. He suggests that North-South bargaining is going on all the time and that to look at the bargaining over the New International Economic Order is to miss the progress that is being made globally in all sorts of institutions.

One can both agree and disagree fundamentally with him. Recalling Gerry Helleiner's remarks, short-term issues are being dealt with continually; long-term issues are not.

The NIEO negotiations are quite novel in the sense that they raise a fundamental set of distributional issues as they have perhaps never been raised internationally before. I am not arguing for or against the context in which those distributional issues are being discussed. Paul Streeten's observations on interstate relations, intrastate relations, people-to-people relations, and so on, suggest how complex the distributional and/or "equity" issues can become.

But the kind of progress that, as Sid Weintraub rightfully points out, is going on and continues to go on all the time is of a different nature. That is, it takes place in a set of institutions, negotiations, and norms which govern interstate behavior and which have existed with considerable influence ever since the end of World War II.

The issue for the Southern countries, when one thinks about a longer historical perspective, is to alter these very institutions, rules, and norms, and thus to alter the payoff matrix of the whole international economic system.

We're talking about distribution of economic gains, of economic opportunities, of political influence, and of a number of other things. In that sense, we could have a great deal of success in the kinds of negotiations that Dr. Weintraub has mentioned, with continued deadlock over the issues that the South is trying to raise in New

York.

That will concern some people sympathetic to Southern goals. But I suggest that the concern or lack of it related directly to a value judgment regarding the fundamental nature of the distributional questions that are, in their very "globality" and scope, novel in the NIEO debate.

Chairman Mills: Thank you. John Sewell.

John Sewell: I am somewhat relieved by this discussion. I was afraid that it was going to be one of those meetings, in the words of Dudley Sears, where the Southerners speak for the South and the Northerners speak for the South.

Not being an economist, I return to a much more pragmatic political question, one having to do with the politics of negotiation and developing countries' perceptions of their political interests and the similar perceptions of the industrial countries.

The process of bargaining has been going on, and I would agree that there's been a great deal of economic progress in the world. I would not agree that there has been very much progress in global negotiations, either in the narrow UN definition or in a host of other international economic issues, including those that we have discussed in a functional sense.

When one looks at the last round of trade negotiations, one takes at face value the decision of most developing countries that they were not very satisfactory. They haven't signed them.

If one looks at the breakdown of negotiations on grain reserves, one sees much the same pattern. The industrial countries try to sort out things among themselves and only bring in the other participants somewhere later in the process.

The most fascinating element in these negotiations, from the standpoint of politics, is why they are taking place at all. I realize that there is a certain imperative built into the various processes such as the UN. After all, we are there and we will remain there.

But why are we in the North concerned with global negotiations? Why do we continue to go back to the bargaining table? Why is President Reagan agreeing, much to my amazement, to attend a two-day summit of North-South leaders in October of 1981 in Cancún?

Clearly, either we are misleading ourselves about our interests or else there is an imperative to return to the negotiating table. When one looks at parallel or similar negotiations, there are two major

45

ones that stick out as negotiations in which the issues have been equally complex and the number of participants equally great.

One such negotiation is the set of arrangements between the European Economic Community and the so-called Associated States that resulted in the Lomé I and Lomé II agreements. In this series of negotiations, fifty-two of the world's poorest countries — and supposedly not very well equipped with negotiating ability — negotiated a kind of New International Economic Order with trade concessions, compensatory financing for commodities, and a whole range of other international measures.

The second striking negotiation, which rarely comes up in academic discussions of North-South problems, is that concerning the law of the seas. Until several weeks ago, these meetings seemed to be moving toward a successful conclusion. A large number of countries were involved, in all, about 164. They were discussing very tough issues, not only economic issues, but a range of political and military/strategic issues. Until the United States decided to reassess its position, we seemed to have arrived at an agreement.

It would be very interesting to know why those negotiations did much better than the negotiations on the common fund or on the NIEO.

One of the problems for the negotiations is the dichotomy between the long run and the short run, about which I think Gerry Helleiner is quite right. From our philosophical standpoint, we in the North tend to want to deal with short-run problems, whereas other people, particularly in the developing countries, see this period as one in which long-run structural problems of concern to them can finally be successfully pressed against the North. The long-run shift is essentially political, and the economic aspects of the negotiations are in many ways a manifestation of the political desire to share power and benefits.

One can argue about who gets what share and what one does about the very pressing problem that Paul Streeten raised of transcending the nation state, but this is essentially a political matter. The question of whether the developed countries are willing to share power in a variety of ways is the central one.

Someone pointed out to me quite recently that, on the tricky issue of debt rescheduling, the Northern countries have for years maintained that all international financial issues should be decided within the IMF or the nexus of the IMF and the World Bank, for that is what those institutions were created to do.

A long-standing demand of the developing countries is for some different way of rescheduling debt, both official debt and commercial bank debt. I gather that the developing countries in the Group of 77 have agreed that this new mechanism and this new process should be set up under the auspices of the World Bank and the IMF, and have decided this quite recently. But as far as I can tell, there has been a roaring silence on the part of industrial countries, despite having this mandate within the context of the institutions on which they put so much stress.

The other dichotomy is that we do face both short-run and long-run problems. It's quite true that the world worked quite decently from 1950 until the early 1970s for both the rich and a good number of the poor countries. The question is whether, in the 1980s and 1990s, it's going to work very well for anybody. It may work very well for Mexico and it may work well for the OPEC countries and certain other countries. It's not working very well for the rich countries at the moment.

One should not be looking at North-South negotiations simply from the standpoint of short-term United States interests, but also from the standpoints of long-term U.S. interests and even broader OECD interests. One of the striking omissions has been the lack of any *Northern* NIEO, if I can put it that way — any comprehensive statement about where we want to end up and what changes we want to achieve given a sharper definition of our own long-term interests.

Chairman Mills: Thank you. Ed Hamilton.

Ed Hamilton: First, I would argue that although Walt Rostow is perfectly correct in emphasizing the need for historical perspective, there is a real case here for emphasizing historical discontinuity. That is to say, one might argue that history is extremely interesting in understanding the past and rather less interesting in understanding the future. We face a future that is quite different, and its primary hallmark is stagflation.

This stagflation problem in the developed countries is really rather more serious than is suggested by the notion that if North and South could simply get together in a room and take it on seriously, we could solve it. We have rather less in the way of tools or knowledge or understanding than such a notion implies. If one assumes that the problem is more intractable than that, then the contextual politics in the developed countries may be quite different from anything that we have experienced in the last thirty years.

Consider, for example, a situation in which the United States Con-

47

gress reduces federal expenditures at the rate that is suggested by current proposals, and continues in that mode for some period of years while the average real income in the United States, and to some degree in other countries, drops. In the context of that situation, if we compare a debate about what has always been and probably will continue to be a relatively derivative subject — i.e., North-South relations — with the main foreign and domestic concerns of the countries involved, we might get a rather different picture of North-South relations in the 1980s.

Second, as for the regional emphasis in Walt Rostow's comments, there is no question in my mind that in this context of greater political difficulty in the North, the notion that one can take on global bargains, or social contracts, recedes as a politically feasible proposition. That one can organize functional or sectoral discussions that are driven by a form of self-interest becomes a substantially more likely prospect.

So the tendency of events is very likely to lead in Professor Rostow's direction. But one must recognize that it is a different bargaining political problem, quite apart from economics, to succeed within a framework or a mosaic of separate negotiations on separate sectoral subjects with separate regional partners who have different interests and involvements in various resource issues.

To get that process to work in such a way that the entire system doesn't dissolve into regional blocs is a very considerable feat, and one that we probably haven't thought enough about.

Third, Sid Weintraub pointed out that people who were seriously negotiating questions of short-term interest had their self-interest in mind and therefore couldn't really think about the overall development problem. It seems to me that this is exactly the point — that self-interest tends to be perceived in necessarily short-run terms. At any negotiation on any specific issue, it is difficult to make any kind of generalized point about North-South relations.

In the middle of a negotiation quite cosmic in content — the Law of the Sea Conference isn't a bad example — an attempt to take into account the general mosaic of North-South relations is extremely difficult, regardless of who it comes from or what the context is. It simply betokens a view that is generally perceived by all the participants, not simply the Northern ones, as one that is out of kilter and irrelevant to the hard-nosed bargaining that is going on.

It seems to me that this is a definition of the problem, not just an observation of the nature of human beings, though it is that, too. It is only when we invest the long-term North-South relationship with the notion of self-interest that we are likely to move in the direction we're talking about here.

Finally, it seems to me Paul Streeten asked the right three questions on the issue of the moral or "solidarity" base for the North-South development objectives. But there are two others that are of equal importance. The first concerns the welfare philosophy: the idea that the world is a community in which there is some obligation to the poor members of the community.

Is it possible, within any reasonably describable set of circumstances and within a meaningful period of time, to make a difference that is sufficiently large so as to register on the scale of decency in the North? Let us assume that we have successfully resolved all of the moral and other issues. If we then put X resources into Y country through Z channel, will the people who are debating the resource allocation question regard the results of these measures as significant enough to make the whole enterprise worth doing?

Suppose, for example, you could raise a country's per capita income from $150 to $200 over a period of fifty years through a very large aid program. It would be very hard to sell that proposition in a Northern political forum because the end result does not seem worth fifty years of effort at the required level of financial support. It is a matter of how well off you are at $200 a year as distinguished from $150 a year in the eye of the beholder — in this case, the Northern beholder.

The second question is whether humanitarian-based aid distinguishes, and to what degree, between short-term symptomatic relief and long-term homegrown relief. The classic example, of course, is food. Is it really doing a country a favor to shower it with large amounts of concessional-term food? Over the long run, does that produce a greater humanitarian problem (i.e., lower levels of domestic food production) than the short-run humanitarian problems being addressed?

It seems to me these two questions need to be added to those posed by Streeten in order to complete the list of the questions that must be asked by anyone who is arguing the moral issue.

One brief and final comment. The issue of organization of the North-South dialogue has been around, of course, for fifteen years, but ideas concerning organization began to take form in the period before the Pearson Commission, during the period of the late 1960s. The one thing all parties agreed upon was that the action should not end up in New York.

The worst thing that could happen would be to have had it end up in the Economic and Social Council (ECOSOC) of the UN or involve the

Group of 77 for exactly the reasons that Joan Spero has put forward: the lack of appropriate mechanisms for negotiation. One could not imagine the North-South developmental issues being successfully negotiated in the UN. So the notion of the process as a barrier under these circumstances is both correct and entirely expected.

Chairman Mills: Thank you. Professor Rostow.

Walt Rostow: I have a few words to say about morality and solidarity, the exchange between Hansen and Weintraub, the distinction between the long run and the short run, and stagflation.

If you look at the motives that operate in the Northern countries in support of foreign aid, you would be quite unrealistic to rule out morality and even religion as one of the factors.

I'm now looking, for example, at the motives that led to the setting up of sustained aid for India and Pakistan in the late 1950s and in 1960-61. What you've got is a large set of converging motives: the India-China competition, Soviet aid programs, and the fact that India was trying democracy. But there was a strand of simple moral impulse to help a lot of poor folks. There was also intellectual excitement at seeing what you could do to help develop so vast a nation. It was a complex mix, and it remains so. Therefore, we should not try to overanalyze exactly what the motives are. We should be conscious of them. The important thing is to get a convergence of motives that will lead to action.

The explicitness and the correctness with which Mahbub ul Haq analyzed the market imperfections issue illustrates part of the difficulty. Suppose the South approaches the North and says, "You don't read Galbraith and you don't understand about market imperfections. We're going to get some countervailing power. Furthermore, we're going to mobilize and take some of that money away from you. Besides, you owe it to us morally." That's an awfully poor way to go about trying to get anybody to help you and to work with you. The important thing to say is, "Here's the problem and, whatever your motives, let's get on with solving it, for we both have an interest in its solution."

Now, this relates to the Streeten notion of solidarity. I worked in the United Nations Economic Commission for Europe during the worst part of the Cold War. We had in that operation the best morale I think I've ever seen in an organization. All of us regarded the goals of European recovery as good, and it led to all kinds of human solidarity. We were doing a job together in which we believed. Don't forget that doing something as pedestrian as setting up the European coal and steel community was motivated by a desire to end, once and for all, the Franco-German confrontation in Europe and to put Germany in a position of collegiality that the rest of Europe could be comfortable with.

So, I don't really buy the view that one has to be explicitly moral and clarify all of these big philosophical issues. The highest sense of human connection and of solidarity with other human beings comes from a translation of moral purposes into something that's recognizable. It comes not by being explicit but by doing things together that matter.

Now, on Hansen's concept of a payoff matrix. The New Economic Order is serious business. But all he wants to do is get production up. There is something missing here in social science terms.

If you look at a cross-section of the growth paths in development, both historically and at the present time, they follow an S-shaped path. That is to say, the very poor countries have low rates of growth that accelerate and decelerate as real income per capita rises. It was to dramatize this pattern that I called my recent book of rather solemn essays *Why the Poor Get Richer and the Rich Slow Down*. We in the North can marginally assist Southern growth, but it has to be accomplished mainly by the efforts of the developing countries. And they must get themselves onto that accelerating path, where many of them now are, incidentally.

There is going to be a shift of power in the world. In fact, none of the real shifts in power have come about through United Nations negotiations or changes in the voting structure in the World Bank. The shifts happened because some of the developing countries have been growing faster than the developed countries.

Brazil and Mexico, for example, have greater stature in the world by any measurement, not because they are members of some kind of club of 77, but because they have become in every sense more competent, more effective nations. Mexico's growth rate has been stable and high, higher than that of the United States, since about 1940.

This is how, in fact, the New Economic Order is going to be brought about. And these shifts in power will come with shifts in capacity. Look at the case of Taiwan. Despite all the complexities it faces, it has as much control over its destiny as any nation in the world. The European banks are all pouring into Taiwan and setting up offices because Taiwan has become a highly competitive nation of 20 million people.

As for the long run and the short run, I do not accept the distinction as useful. The long run is simply the accumulation of what you do in the short run. If you don't ever start doing something in the short run, you don't get a long-run effect. Some things you do in the short run for short-run purposes have very important long-run consequences. For example, the setting up of the World Bank and the Inter-American Development Bank. People take them for granted, but these are extraordinarily powerful long-run institutions. Where you end up results from the accumulation of the things you do from day to day, and some of those things cast long shadows.

Gerald Helleiner: In both directions.

John Sewell: Including things you don't do.

Walt Rostow: Exactly. The things you don't do also cast long shadows. I think that Keynes was quite wrong when he said that in the long run we're all dead. I would say the long run is with us every day of our lives and that the actions you take or don't take shape the long run.

Finally, Ed, on stagflation. Of course, you're not going to solve it just by exhortation. But before you leave, I'll give you an explicit proposal.

Ed Hamilton: You don't mind if I pass it on, do you?

Chairman Mills: Jorge Navarrete.

Jorge Navarrete: I would like to comment briefly on the approaches to the agenda that were mentioned in several of the interventions.

For me, it is sometimes difficult to understand the criticism of an agenda in the global realm and the support for alternative agendas which are more or less similar to the "global" one that is so often criticized. In the global realm, we have the five areas of discussion that were mentioned by Joan Spero. And in the alternative agendas that have been suggested, most of those areas of discussion are also present. Some of them are narrowed a little bit, but they still remain very wide. To mention food and agriculture, for instance, is to mention an area as wide as "raw materials" or as wide as "development" in the global negotiations approach. Everybody mentions energy, both in the global realm and for the other purpose.

But the real problem is to define what elements are within these very wide areas of concern on which everybody more or less agrees. This is the problem in New York in launching the Global Negotiation and it will be the problem everywhere for meaningfully discussing these questions.

To approach this kind of problem, I think we need a combination of the sense of urgency in the short term and of attention to long-term progress. For instance, it would be insufficient to discuss food as merely a matter of food security. We must also do something important in the area of agricultural development in the developing countries. Likewise, it will not be enough to discuss protectionism if you are not prepared to discuss long-term structural adjustment, that is, the industrialization of developing countries and the reindustrialization of developed countries.

Sometimes pressing short-term problems are the consequences of structural problems that can be resolved only within the context of a long-term time frame. Professor Rostow mentioned that the decision to establish the World Bank and the Inter-American Development Bank has had long-term consequences. But this is precisely the kind of decision that is taken not simply to respond to short-term necessity.

Chairman Mills: Mahbub ul Haq.

Mahbub ul Haq: I was at first greatly impressed by the very constructive and pragmatic mood that prevailed in this discussion. But I also started getting a bit worried about the degree of humanity around this table, which could prevail only if we were ducking the real issues. I would, therefore, like to emphasize a few of the crucial differences on the agenda, the approach, and the process of negotiation, which our exchange may have obfuscated.

It's very easy to agree on short-term issues of food, energy, recycling, and protectionism.

Walt Rostow: What makes you think that's short-term?

Mahbub ul Haq: It is short-term in the sense that these are the most immediate and urgent problems that developing countries of the world face. But I distinguish between short-term and long-term areas, the latter involving major structural changes between our nations and international institutions — the International Monetary Fund, the World Bank, and the other Bretton Woods structures. The long-term issues, which are also part of the negotiating process, are the major structural reforms that the Third World is seeking. And there is a very direct link between the two.

I agree with Mr. Rostow that the distinction is serious because each of the short-term issues like energy and food has very long-term structural implications. The differences in approach come as we negotiate each one of them. Let me take up the subject of energy, just for illustration.

It is very easy to agree that we need to build a new basis for energy development in developing countries, although we may get an argument from OPEC. After agreeing on this objective, we move to the kinds of energy plans that developing countries have to prepare, such as pricing policies and conservation measures. A lot of this will be determined in national elections and some of it perhaps in regional elections. But the global dimension comes in as you start looking for the very large resources needed for investment. Eighty billion dollars a year is a very substantial investment in energy in the oil-importing developing countries, especially if these are to continue with their other development efforts, including the required growth in employment levels, during this phase of adjustment.

Now, where does the conflict in views enter? It's not on the size of the investment, for developing countries are quite prepared to accept world estimates that have been prepared on it. The key and contentious question is what kind of international or global institution is to be established to achieve these energy investment goals? The World Bank proposed the formation of an energy affiliate that would· channel about $5 million a year into such investment and would cofinance it with private oil companies.

Immediately, a struggle begins. OPEC, which is willing to contribute a major part of the funds needed for this energy affiliate, will not do so unless there is a different balance of power in the voting structure, in the management structure, and in the separation of stock.

The OECD countries have said they will not participate unless the affiliate is integrated with existing stock and with the existing management of the Bank. So there is an immediate conflict between the short-term response on which there may be a very generalized agreement and the long-term changes that the developing countries seek. One might well advise them to take the money at this stage and to leave the question of new institutional devices to the future. Obviously, it is not that simple.

In the context of each proposal, whether it concerns energy, or food security, or recycling, the issue of new devices for existing international institutions arises. Do they alter the existing balance of power and influence? Do they reflect realities which have already changed and which may change much more in the future? Or do they try to recapture the past? This becomes a very major issue in the North-South dialogue. And as the current energy dialogue demonstrates, negotiations break down — unfortunately, in this case — regarding an issue on which developed countries, OPEC, and oil-importing developing countries are in total agreement on "short-term" needs.

Furthermore, ideology also enters the picture. Last year, the United States, which initially supported the energy affiliate proposal, began to back away from it, suggesting that international institutions like the World Bank should not borrow from the market to lend to developing countries — that oil companies should do it directly. But oil companies can't do it directly, for there is too much risk. Some international institution has to play a catalyst role in predevelopment and exploration, which are risky endeavors.

Moreover, developing countries don't want assistance on that basis because many of them are afraid of the unequal relationships that might result from having to deal directly with the major oil companies. So we quickly come to an impasse, although we thought we had very considerable agreement when we started.

John Sewell mentioned the problem of process. Process is indeed a major problem, and the question is how to solve it. If an energy affiliate is negotiated within the World Bank structure, certain pressures immediately emerge reflecting different perspectives.

In the New York UN setting, the Group of 77 said, "Let us have this energy affiliate under the UN umbrella and have negotiations regarding it here." That's too unwieldy, and we need something in between. Each time a certain proposal comes up, we could perhaps have a special committee that would deal with the proposal within the UN framework, where political pressures are always in the background. So even though you can't negotiate in the UN Committee of the Whole, you might be able to negotiate in smaller, special UN committees established for this purpose.

Alternatively, proposals could be negotiated in the Development Committee, which consists of twenty of the most powerful finance ministers in the world. The credentials of the Development Committee are very low; it was set up in 1974 and hasn't done anything in seven years. But perhaps this new function might provide it with a renewed sense of purpose. Should developing countries take a chance in that forum? It takes a tremendous amount of courage for them to overcome their present cynicism and accept any institutional answer short of the UN, where they know they can't be outvoted. In sum, the process has to be thought through very carefully.

Finally, we must face the issue that even where there are common interests in the long run, there are very heavy short-term costs of adjustment for developed countries as well as developing countries — and even for the OPEC countries as they strive to translate the new-found oil wealth into real sources of growth.

It's politically expedient to postpone these adjustments and their costs through various means. These costs accumulate over time, as energy adjustment, environmental adjustment, and food adjustment have in developing countries. Population problems would also have accumulated, except that now some action is being taken. These costs accumulate rapidly, and the problems causing them become very difficult to deal with the longer the effort is postponed. So one must analyze carefully the costs of adjustment in the short term vis-à-vis the long-term benefits, and consider how to make this cost of adjustment palatable to those interest groups that will otherwise attempt to block the longer-term adjustments.

All this amounts to a very difficult exercise in political leadership. People who are up for election normally like to rescue the next election rather than the next generation.

I do not have ready answers to this dilemma of reconciling short-term costs of adjustments with general long-term mutual interests. Perhaps you use whatever bargaining power and whatever arguments are available in all forums. Maybe, as Professor Rostow argues, one shouldn't make real and potential conflicts of interest between North and South too explicit. I made them explicit, Professor Rostow, because you raised the issue last night and I thought I owed you a reply.

Chairman Mills: Thank you. Joan Spero.

Joan Spero: I would like to conclude on a note of pragmatic pessimism on the one hand and optimistic idealism on the other hand.

I think Gerry Helleiner raised a crucial point relating to this short-term versus long-term issue. And let me say from my perspective of watching the process, not of the UN but of the United States government, that I am very pessimistic as to whether governments today are willing to focus even on the short-term aspects of the North-South dialogue, let alone the long-term aspects.

Mahbub has given us a clear example of a case in which everyone agrees that energy development for developing countries is a crucial issue, both for the developed and the developing countries, and yet the United States has done a complete about-face on a program which it initially supported in the World Bank. To cite another example, you find people in the Development Committee and in Northern treasuries denying that the problem of recycling exists, and you find the OPEC countries taking the same position.

So, I'm very pessimistic about the short run, and I think there are a variety of reasons for the current situation. An obvious one is that the present economic situation — stagflation — is the primary issue for economists and politicians in the developed countries today.

The simple efforts in this past year to get through the Congress a replenishment of IDA or new funds for the IMF, let alone for the World Bank, met with great difficulty. The domestic economic problems are the main focus of decisionmakers today, and they're not thinking much about the North-South problem.

Furthermore, I don't think that they even recognize that there are some areas of mutual self-interest. Wherever they do recognize some kind of self-interest in terms of export markets or bank lending, that self-interest focuses on a very small number of developing countries. The rest really don't matter to policymakers.

So the New International Economic Order is vulnerable on the left and on the right. It's vulnerable on the right for all of the reasons I've mentioned and because of the traditional economic viewpoint that generally rejects the entire Southern critique of the present system. It's vulnerable on the left because of its relation to basic human needs. Are you transferring funds from the poor in the developed countries to the rich in the developing countries? And I think that's a fundamental dilemma, not only in the conception of the New International Economic Order but in its political feasibility in the developed countries.

There is thus a real problem not only in the long term but also in the short term. In the past, much creative thinking on long-term problems for developing countries — for example, the creation of the Inter-American Development Bank — didn't reflect an effort to solve economic problems. Establishing this institution was a response to political problems. Perhaps the best thing we can have is serious crisis in El Salvador, for it may lead to a new injection of funds into the Inter-American Development Bank. With respect to North-South relations, I don't think that the policymakers respond to economic crises. They respond to political crises.

So much for pessimism. Now let me share some of my optimistic idealism. I was provoked by all of the criticism of the United Nations, in which I share and which I initiated in this discussion.

I believe the UN in New York still performs an important function — to keep the pressure on by keeping the issues of importance to the developing countries alive. It is, if you will, a visible pressure cooker into which the developed countries sometimes are willing to put themselves. Interestingly, OPEC is not willing to put itself into that pressure cooker yet.

If global negotiations take place, they will have an impact on the North-South dialogue in other forums. For example, the changes that were made recently in the percentage of borrowing allowable at the IMF were perhaps partly due to a consideration of what might happen in the Global Negotiation.

The UN also makes bureaucracies in the developed countries pay attention to issues which wouldn't concern them otherwise. There are certain bureaucrats who have to think about these things because their bosses have to go before the UN prepared to speak and respond. That may be true in the developing countries as well as in the developed countries.

Chairman Mills: Thank you. Gerry Helleiner.

Gerald Helleiner: Here we are in the early months of a new United States administration which has new views — on these issues as well as others — that are being much discussed all around the world, and we're in a session that is called "The Setting." I would dearly love to have someone who knows much more about it than many of us here do, express some views about the prospects in this administration.

There has been anxiety, certainly long before this administration came along, about tendencies within the OECD to circle the wagons, to form "Fortress OECD." The fear is that, if things get too uncomfortable in the UN organizations, the North will, at least *de facto*, "drop out."

So I would like to hear, for example, someone discuss what we can expect from the United States in the trade sphere. There are those who say that ideologically the prospects for freer trade are better than they were before the recent election. I've heard similar arguments on the recycling issue, but on other grounds it may be more difficult to find room for optimism.

Chairman Mills: I can advance a very brief comment. I'm not in the business of being pessimistic, because to be pessimistic is to admit to the other side that you have given up. And none of us can give up because we are all trapped in the same situation. We are victims of an inexorable process, and if we miss opportunities for progress today, we'll have to face greater problems tomorrow under much more rigorous circumstances. That, after all, is the lesson of the last seven or eight years, since the proposals on the New International Economic Order were presented. We have fooled around with certain issues, and to an amazing extent we keep coming back to them because so little has been accomplished.

For example, we discuss international monetary reform, we have in fact for just about ten years, but we keep skirting deep differences and important issues involved. The world might have been a little better off today if we had had the guts, over the last seven years, to make some serious and fundamental decisions. Instead we witness a deepening crisis in the global economy.

I don't see much to hope for in terms of what the United States administration might do today. But those of us who are interested in getting something done will have to find a means of overcoming the difficulties caused by changes in administration — whether in Jamaica, in the United States, or anywhere else. The world cannot wait in its efforts to deal with really crucial issues until each administration has come around to making its decision. On such matters we have to establish a deeper commitment and a wider consensus.

Countries have a problem of protecting their interests in the world in spite of the internal organization of their political systems. This is something all of us should be concerned about, for it's not only the United States which has this problem. But if you look back over the record, not only the UN record but elsewhere, the excuses that have been given by different countries for not taking action, especially on critical economic issues, would fill volumes. Always there is a reason for not making a decision now. Things are either too good, or they're too bad, or they're too uncertain.

I believe that historians will look back and chastise our generation for gross irresponsibility in the face of very critical problems. The people of thirty-five years ago, faced with very serious global problems and dangers, came up with institutional and other far-reaching innovations to deal with them.

What we are trying to do today in the matter of establishing new global economic relationships is not at all easy, and the blame for failure cannot all be placed on one side. But what is most remarkable is the extent to which we have gone around in circles. So somehow we have trapped ourselves into this soft talk, determined not to quarrel with each other as we did at the Sixth Special Session of the UN General Assembly in 1974. We must now talk politely and without criticism!

It's really a very, very serious matter, and that's why those of us in the UN need to reach out to those who are outside. Not that all of those who are outside necessarily have done much about the process; some have, some have not. The issue of North-South economic relations has been ignored in some circles, in the Third World as well as in the North.

It's not an issue that should be confined in the way it is. There is an inadequacy in the political process involved in such a global issue, and until we make links between the UN system, national preoccupations, national political processes, and the academic world, we will make little progress. We have to build some sort of political network in the widest sense of the term. This network is needed to make the issue real and alive to a constituency that goes well beyond the "converted" members of the Group of 77 and those in the North who are involved and interested. It takes a combination of interests to make the issue a reality, to make the discussion a reality, to make differences clear, and then to move to constructive negotiation.

Walt Rostow: I think that what Joan Spero said a moment ago is the proper reply to Mahbub's question, and Jean Monnet expressed a similar idea when he said that governments only change policy when they are forced to, and they're usually forced to only in the face of crisis.

The Inter-American Development Bank was indeed the product of a political crisis: when the vice president went to Venezuela, a lot of rocks were thrown at his car and United States aircraft carriers were sent South. In the midst of this incident, the president of Brazil sent President Eisenhower a letter. He understood that at last President Eisenhower, having almost lost his vice president, was going to listen to his brother, Milton Eisenhower, who had been campaigning for aid to Latin America. He proposed Operation Pan America.

What lay behind Operation Pan America and what lay behind the long battle for foreign aid in the 1950s was a North-South consensus — which does not exist now and which it's the duty of a group like this to create.

We who care about these matters and have committed a large part of our lives to them must not merely echo in either polite or noisy or amusing ways the sterile nonsense that's been going on, with the South making its complaints to the North and the North managing to fend them off more or less skillfully.

If we can't come to an honest working consensus of the kind we had on foreign aid in Latin America and India in the 1950s — which embraced intellectuals and people who care in the North and the South — then the politicians won't listen to us in the moment of crisis.

At a conference in Mexico held in the wake of the OPEC price decision, friends with whom I had worked a good part of my life in the South suddenly became quite excited about what this action might mean more broadly for extracting resources from the North. I felt sad because in the long run the high prices would hurt the developing countries and, in addition, would divide those people who had worked rather effectively together on behalf of the developing world in the North in the previous generation.

So we may have to wait until a more acute crisis concentrates the minds of the governments, in Dr. Johnson's phrase, but we also have a duty to build a consensus among ourselves that does not quite yet exist.

Chairman Mills: I will make a few comments to summarize this very interesting session.

There has been a good discussion on the justification for the New International Economic Order, but we are still in the stage of misunderstanding.

The Sixth Special Session of the UN, held in 1974, formally launched the NIEO issue, and in 1975 we had another special session. In between the two, we really agonized about how we could talk to each other, since the 1974 session had been seen as quite acrimonious. The Seventh Special Session was very polite by comparison, and we all thought it was a success because we talked mildly to each other, and we reached a "consensus" decision at the end that was supposed to provide the foundations for real negotiations. In retrospect, it was really a failure, a conspiracy of softness. We should understand how and why this happened. But we have not learned the lesson, and therefore we have continued to delude ourselves.

We have not reached a broad consensus on the need for change in global economic structures and relationships in more than seven years of discussing the issue. There has been talk of consensus. There has been a convergence of rhetoric: I can quote you statements made by leaders of Western countries that suggest a real degree of convergence. But there is no consensus on purpose and on action.

It's impossible to believe anything else when you see how little progress we've made in the negotiations at the UN. So let's face the fact. A basic philosophical change must first take place, not a change covering the details of what the Group of 77 will prescribe. If we pay too little attention to the need for this philosophical change, it will be very difficult to make any significant progress in the negotiations.

Joan Spero has talked about process. It is of the first importance. It is a major obstacle. I associate it with the remarks about pragmatic approaches and short-term and long-term approaches. And what the "pragmatic" and "short-term" approaches translate into most often is the following: when we are faced with complex problems and comprehensive proposals, let's take the easiest things first, those that we can agree on, and leave the more difficult things until later.

My personal view is that this approach is unacceptable; it will not work. It's not that there isn't some element of logic to it. But if one group of people make proposals, some of which are far-reaching in terms of changing the relationship with others, don't ask them to put off these complex proposals and to deal with those that are easily managed, the short-term ones, which bypass the issue of altering relationships of power and influence. The same holds for international negotiations.

How do you then face the reality that it is easier to make progress on certain matters more readily than others? How do you, without rejecting that approach, not trap yourself into forever dropping the long-term issues? This question came up in preparations for the Seventh Special Session at the UN, in the Committee of the Whole, and at the CIEC meetings in Paris; it comes up every time you try to promote or renew the dialogue. We are constantly faced with the proposition of starting with the easiest things, or developing an emergency program.

As for the UN itself: we've heard some of the comments about how the UN is seen in some circles today, and much of it has some truth. But I would add a correction. For a number of reasons, including the feeling on the part of some that the UN was adequate to the task of dealing with the North-South issue, it was hijacked. It was taken out of the UN to the Conference on International Economic Cooperation (CIEC) in Paris (1975-77), where only twenty-seven countries participated.

What happened? There was much analysis in Paris, and it was a very expensive exercise. My country was there among others. We had among the delegation eminent members of our own community who sat in Paris for eighteen months. Some very good work was done, but they failed to resolve a single major issue relating to structural changes on the agenda. They tried to take on a political issue, at least an issue which has considerable international political implications outside the political forum. When the UN fails on the same issue, the fault clearly does not lie within the institution and its processes alone.

The UN is very imperfect, and our duty should be to try to improve it. Comments have been made in this discussion which point us in the right direction. The worst thing that can happen is for those of us in the UN to pretend that we are doing something which is impossible for us to do and which should be done elsewhere. How we define this link between the competence of the UN and the competence of other bodies is one of the major challenges that we face here.

We have heard discussion supporting the "pragmatic approach," whereby the Brazils and the Mexicos pass through stages of development and join the high orders of mankind. Of course, this process is occurring, if slowly. But if you left it to that process, you would always have a Third World, since the great income and other gaps which separate the richest from the poorest nations would never close. In the moral sense and the cultural sense, this is what the "pragmatic approach" is about.

During the decolonization process, a period of about fifteen to twenty years saw the freeing of eighty to one hundred countries. If this process had been the subject of global negotiations in the UN, they would still be trying to decide the agenda and the mechanism for debate and discussion.

Session II
Energy and the North-South Impasse: Three Perspectives

Chairman Weintraub: What's been happening in the world energy picture is well known. The shift in the world oil price which took place in 1973 and 1974 was truly one of those moments of revolution. In the world economic structure, changes have taken place which we're still trying to live with.

The recycling issue was made quite critical by that particular change. The resource transfers that went from the oil-importing countries to the oil-exporting countries in most years dwarfed those which went for official development assistance to the poor countries. We've had major shifts in financial transfers, and in due course they're going to translate into shifts in real resources from oil-importing countries to oil-exporting countries.

We have also seen major shifts in power. Power shifts take place not because people talk about "political will" and pass resolutions in international institutions about who should have power; these shifts occur because countries take action in everyday events to produce these power changes. The OPEC pricing change is one example of a major power shift which is then reflected in institutions, in actions, and in the things we do.

There has been some mention of the problems of OPEC cohesion with the other members of the Group of 77 in various forums. One of the manifestations of cooperation among these groups was the facility through which Mexico and Venezuela provided assistance to the countries of Central America and the Caribbean in the purchase of oil. Perhaps this is a harbinger of other similar activities by oil-exporting countries to help oil-importing developing countries.

Let me now introduce Jorge Eduardo Navarrete, our first speaker.

A View from the Oil-Exporting Developing World

Jorge Navarrete: It would clearly be presumptuous on my part to attempt to present the international energy issues from the point of view of the oil-exporting developing countries as if a common point of view had been defined or elaborated by the oil-exporting developing countries or by the members of OPEC. It has not.

It seems advisable in this situation to narrow the scope of my presentation. I will do my best to offer you the point of view of one particular oil-exporting developing country on international energy matters as well as on the problems of international energy discussions.

This precision is especially needed because my country is still not always seen as an oil-exporting developing country. In a very interesting and useful recent paper called "North-South, A Fourteen-Point Action Program," authored by a very prominent British political personality, Mexico is presented as one of the nonoil developing countries deeply in debt. Leaving aside for a moment the debt situation, it is clearly misleading to think of Mexico today as a non–oil-exporting country.

Mexico's oil-exporting operations resumed significantly only two or three years ago, and that perhaps explains the persistence of this erroneous view. On the other hand, it is equally misleading to see Mexico as simply an oil well or as just another oil-exporting country.

I come from the only country so far which has presented a formal proposal for dealing with the global energy problem of today through worldwide negotiations. I am also a government official of that country. Of course, I am not here to present the official view of my government in these questions. As is usually said, I am participating in my "personal capacity," whatever that means.

My purpose is to present the main issues and then allow the discussion to clarify and deepen our understanding of these issues.

In considering the role of energy in the North-South impasse, one question nearly overshadows all the others: the question of the approach to international energy issues in multilateral economic discussions and negotiations. It is clear by now that for more than five years the international community has been unable to find an approach to energy questions that is universally acceptable as well as practical and workable.

As a consequence, significant multilateral negotiations on the subject of energy have never really begun. In this respect, it is worth remembering that CIEC, the Paris conference from 1975 to 1977, was initially conceived in the North as an attempt to discuss only international energy questions. We know from its history that CIEC would never have taken place if this initial approach had not been dropped and a wider one adopted. In a certain sense, then, the roots of the "global negotiations" notion and approach to North-South issues are to be found precisely in this Northern effort in the mid 1970s to isolate energy from other problems of the international economy. On the insistence of the developing countries, the CIEC agenda was broadened and the conference itself organized into four interrelated committees or commissions dealing with energy, raw materials, development, and finance.

Five years later, there is still no agreement on how to organize a round of global negotiations on precisely the same subject matter. But at least the global approach seems to have been universally accepted and in a form which used to be considered a particularly prestigious one, consensus resolutions of the United Nations General Assembly.

So it is important to ask why, one and a half years after the adoption of these resolutions, practical results have not been achieved. I suspect that there are several reasons for this.

I suspect, for instance, that some of the countries that joined the consensus for the resolutions in 1979 were not fully convinced of the wisdom and practicability of the global approach. As they see more clearly some of its implications, they are unwilling to let the process begin. Preferring not to openly challenge the concept, they drag their feet as much as they can in discussions of agenda and procedures, maybe in the hope that, since the life expectancy of new ideas and approaches in international organizations is a very short one, the concept of global negotiations will fade with time.

Another reason is perhaps of a more fundamental nature. The global approach was accepted for CIEC in Paris shortly after the first oil shock. And it was also accepted by the UN General Assembly during the second oil shock. In critical situations, the trade-off behind the approach seems more attractive: the South agrees to a discussion of energy if other questions are also included; the North, on its part, agrees reluctantly to negotiate on other issues, if energy is included. Afterward, when the shock has been absorbed and its consequences palliated in some way, the urge to engage in international negotiations on energy is far less powerful. These issues become less important in the face of conditions like the present one, where there is an oversupply of oil, low economic growth, and greater domestic concern over inflation and unemployment in advanced societies.

But there is at least one further reason, and this is perhaps the most important one. It is not an easy task to implement the global approach to international economic problems, including that of energy, in terms of selecting an agenda, identifying priorities among agenda items, and designing procedures which do not undermine the competencies or fields of operation of already existing international institutions. This is particularly difficult in a committee with 150 voices, although admittedly the number of voices effectively heard is much less than that.

It is important to have a very clear understanding of this fundamental question of approach. The global approach, which basically implies not isolating one single issue, particularly energy, is now and will be in the foreseeable future the only acceptable approach to multilateral intergovernmental economic negotiations. This is a political prerequisite reflecting the interrelated nature of the international economy. If we are unable to find practical ways to put this global approach into operation, the stalemate in North-South issues will be a long one indeed.

This same question of approach has a second dimension. Energy means energy, not just oil. International negotiations about energy cannot be concentrated on oil alone, and even less on short-term questions of oil supply and demand. A global approach to the energy question must encompass all sources of energy and such diverse and complex matters as supply, demand, consumption, technology, and investment. It is difficult to foresee meaningful discussions about the conditions of oil supply if there is not a parallel willingness to seriously discuss transfer of nuclear technology, equipment, and materials, to mention but one example.

Besides the matter of approach, it is important to recognize that there has been some progress in the international perception of energy issues. In CIEC itself the Energy Commission was a fruitful exercise; it increased our perception of the very complex nature of the energy transition period in which the entire world economy is involved.

The new international development strategy for the 1980s recently adopted by the UN General Assembly includes, for the first time, two sections on energy — one in the chapter on goals and objectives and the other in the chapter on policy measures. These references to energy are not fully satisfactory to everybody — after all, they are a product of much debate and compromise — but at least they reflect some degree of common concern about the problem and some degree of commitment to shared solutions.

Later this year, in Nairobi, an attempt will be made to formulate a World Action Program of cooperation in new and renewable energy sources. The developing countries are trying to define a meaningful South-South energy cooperation program. All these actions, partial and limited, are important, I think, as steps toward the goal of international cooperation in the field of energy.

The idea behind the Mexico-Venezuela oil agreement for Central

America and the Caribbean is to offer two types of benefits to some of the countries most seriously affected by the international energy situation. The first is a full predictability of supply which eliminates the need for them to go to the spot market and normally incur heavy financial burdens for their imports of oil. The second is to offer some financial relief in the form of soft loans covering 30 percent of the cost of the oil that they are importing from Mexico and from Venezuela. It's not a very complex cooperation scheme, but in the opinion of the countries that put it into operation several months ago, this program is an initial step toward wider and more important arrangements at regional or global levels.

To finish this initial intervention, I think it would be appropriate to mention possible energy issues which could be appropriate items for future North-South discussions. As everybody recognizes, the problems of energy-deficient developing countries have to rank very highly on any kind of North-South global agenda. These problems involve, of course, both short-term financial measures, to respond to pressures of balance of payments, and longer-term measures, in order to develop their own energy resources, both conventional and nonconventional.

A second priority will concern measures to assure a smooth evolution of the transition period. The process of substitution of energy sources will imply that oil and other nonrenewable hydrocarbon energy resources are devoted more and more to higher economic uses.

A third priority will be a freer, more open flow of energy technologies to developing countries as a contribution toward progressively less dependence on imported energy.

A final priority issue concerns predictability. Both developing and developed countries could benefit from a more predictable energy environment in the world economy.

Chairman Weintraub: Now, Ambassador Donald Mills.

A View from the Non–Oil-Exporting Developing World

Donald Mills: In September of 1979, President Lopez-Portillo of Mexico spoke at the United Nations General Assembly. Devoting his entire speech to the question of energy, he proposed a world energy plan. But he also said, early in his speech, "I trust I shall say nothing new. It would be grave indeed if at this point in the crisis there were still something new to be said." I really do not expect to say anything particularly new either, but I hope to offer thoughts which might help to promote a discussion that will produce some new insights.

The fact that the president of Mexico came to the United Nations and spoke in the General Assembly on the subject of energy was a most significant political development, especially when you recall that, in previous years, very little of an explicit nature was uttered about energy in the General Assembly. It was a very sensitive issue. We were all groping for a way of approaching the subject without complicating matters even further. But this General Assembly was significant in that many of the attending ministers spoke about energy. It reflected a change of mood, a movement toward greater realism in approaching the subject in the international forum.

One authority on the subject has said that the energy challenge of each developing country is to find ways of organizing itself to deal effectively with the problem of obtaining enough energy to maintain economic growth while accomplishing the transition to the postpetroleum era.

I would like to suggest that among the many things that developing countries — particularly those that are dependent on imported energy — must do would include the following. They must try to secure supplies of imported oil on the most favorable terms possible. They must try to obtain additional funds for balance-of-payments support to offset increased prices of oil and other imports. Those who have the prospect of finding oil must try to obtain investment funds and technology for oil exploration. They must establish conditions for promoting access to and use of new and renewable sources of energy. To accomplish these goals, they must give impetus to the North-South negotiations and also give increased attention to South-South cooperation.

Each of these approaches contains its own economic and technical challenges, but our interest here is particularly in international political considerations.

Every now and then I come across the term "free world oil supplies," and I wonder who is included today in this category and who is not. But still, the concept of free world oil supplies points to a major political concern in the energy situation. Some of the countries in the North have seen themselves threatened by a possible restriction of supplies, either for political or for other reasons.

Admiral Turner, former director of the CIA, in testifying before the United States Senate Committee on Energy and Natural Resources, said, "Competition for available oil supplies not only will put added pressure on East-West relations; it will strain relations among the industrialized Western nations themselves."

We have not forgotten the measures taken by OPEC countries in 1973. They concentrated the mind wondrously, as we all know. Since then, the revolution in Iran and the Iran-Iraq conflict have demonstrated that internal political difficulties or conflicts between developing countries that are major oil producers can significantly change both the supply and the price of oil.

Developing countries dependent on imported oil have felt for some time that leading industrialized countries have been trying to form a partnership with the OPEC countries. There is clear evidence of this behavior, with which we are all familiar. I'm not criticizing them for it, but merely pointing out that initiatives have been taken at high political levels in the Western world in an attempt to arrive at a far-reaching accommodation with the OPEC countries.

Now, as I say, it is the right of the industrialized countries to do this. But countries like mine would view this with some concern — not only because it might affect our oil supply situation, but also because it might hinder our own attempts to establish an effective partnership with OPEC countries in respect to North-South matters as well as oil supplies.

You can see today the great preoccupation of Western countries with the Persian Gulf political situation and with the Palestinian question. And we can see, of course, the developing relationship — if I might capture it with an innocent phrase — between the United States and Mexico, with oil and energy being a most prominent part of the concern north of the Rio Grande.

In an article some time ago, Jorge Sabato, a nuclear scientist from Argentina and a member of the Club of Rome, argued that the real crisis is not in energy or oil but in a life-style and a production system that have abused a particular resource. According to Sabato, the so-called energy crisis is not a crisis related to the physical availability of energy resources, but a political crisis centered on the struggle for the international control of oil with enormous repercussions in other fields. Certainly, there have been those who have advocated military action in the Persian Gulf area if there was any risk to continued oil supplies to the West.

More recent developments in the Gulf area, if they haven't increased that particular risk, have certainly increased the fears of some people that Western involvement in the area, in a military sense, is not an impossibility.

Developing countries are not in the strongest position to assure themselves of available oil supplies and are, therefore, deeply concerned with international developments which might affect their present degree of access. Partly in response to such concerns, Harland Cleveland has suggested that a "fair access" system should be developed in which, in any crisis, oil-importing developing countries would automatically be assured of a certain amount of oil according to a formula based on the basic oil needs of these countries.

Of course, much depends on what one accepts in terms of the prognosis. There are projections which suggest, as Admiral Turner did, that the market is going to get tighter and tighter. Others suggest that it's going to be quite the opposite and that there will be a dramatic reversal of the situation in so far as the availability of oil supplies is concerned or a great reduction in the reliance on oil for some of the conventional purposes.

But we have seen the effect of Middle Eastern concerns and political problems on the oil situation. Following the oil embargo in 1973 and 1974, some developing countries actually expressed their solidarity with the Arab cause by breaking off diplomatic relations with Israel. And there was certainly an implied connection in some instances between willingness to display that type of solidarity and some form of concessions relating to oil needs of those developing countries. I have the impression that now there is no explicit political link being made between concessions from the OPEC countries and the Middle Eastern situation. Developing countries are now searching for a formula for direct assistance or some other form of cooperation with OPEC members. Over the last eight years, those of us who have been involved both in the UN and elsewhere have been preoccupied with this question.

In general, the developing countries have tried to continue to support the principle of permanent sovereignty over natural resources, which is the principle underlying the action of the OPEC countries. They have sought individually or regionally, and on occasion through the Group of 77, to obtain some general concessions from the oil-exporting countries in the form of rebates, assured oil supplies, or long-term financing assistance arrangements.

We've heard about the arrangements made at the regional level by Mexico and Venezuela. Trinidad and Tobago have come out with a similar facility. But developing countries are finally expressing more and more openly their concern about the impact of oil prices on their economies. OPEC countries have reacted, and the interplay — or call it negotiations — between the oil-importing and the oil-exporting countries has become more and more open and explicit.

OPEC and the oil-exporting countries in general do not wish to find themselves being blamed for all the economic difficulties faced by the developing countries. And they certainly do not wish to have oil treated any differently from, say, wheat or manufactured goods. They don't want oil to be singled out in a special way.

The director general of OPEC, Mr. Shahanta, in an article of last October discussing this relationship between OPEC and the oil-importing developing countries, said, "It's particularly a mistake to exaggerate what OPEC member countries can do to assist other developing countries and to argue that OPEC is the only qualified midwife for the birth of the "New International Economic Order." He points out that the aggregate GNP of all OPEC countries, which amounted to $379 billion in 1978, represents barely 4 percent of the global GNP, 7 percent of that of OECD members, less than 50 percent of that of Japan, and only 17 percent of that of the United States.

Let me also say something about new and renewable sources of energy. As in the case of oil, so it is in the case of new and renewable resources. Developing countries must create the means of monitoring research in areas like solar energy and making sure that they have, individually or collectively, the means of benefiting from these developments. This means, of course, reviewing their national development plans, which, like those of the Western world, have been based on expectations that energy would be cheap. They now realize that their style of development will be seriously affected by what is for them the high price of energy. The Conference on New and Renewable Sources of Energy, which takes place in August of this year, should be of crucial importance to developing countries.

The following is a statement made by Messrs. Howe and Tarrant in an Overseas Development Council publication entitled "An Alternative Road to the Post-Petroleum Era: North-South Cooperation." In talking about new and renewable sources, they have this to say: "It might seem as if the industrialized countries were saying to the Third World, 'You take the sun, the wind and the wastes. We take the oil and the nuclear power.' Now, seen in this light, it would not be illogical for the Third World to conclude that advocating renewable energy amounts to a great conspiracy to keep the world's poorer nations from modernizing." The developing countries didn't say it. I'm quoting.

I can't help recalling the issue which arose in the early stages of the movement toward establishing a full-scale global approach to environmental matters through the UN ten years ago. At that time, there was a fear on the part of a number of developing countries that a strict approach to environmental affairs would limit what developing countries could do in the matter of establishing certain forms of industry.

This created serious difficulty in the early stages of consideration, a difficulty finally overcome. Add to that, though, the fact that in the case of nuclear energy, the concern about proliferation of nuclear capability, particularly for destructive purposes, has led some if not all countries in the West to try to restrict the spread of that capability.

So a major political issue now facing the world is about who shall or shall not be allowed to acquire nuclear capability. Again, you can see the reasons for continuing conflict here. We are witnessing the emergence of a pattern: some of the things that others have done in the past and that they benefited from are now considered taboo because these things are unhealthy. This is a delicate matter as it relates to energy, especially if there continues to be a very tight oil supply.

Developing countries have been very much interested, of course, in having some of the surplus funds, whether from OPEC or elsewhere, flowing to their countries for investment. And the issue of recycling is another sensitive issue. It was much more sensitive in my experience in 1974 and 1975, when recycling meant grabbing hold of this wonderful amount of money that OPEC was getting. You couldn't use the word in certain circles.

Now it can be discussed, and developing countries are anxious to find ways of attracting some of the funds which could be invested in their country. It is also now a subject of discussion among OPEC countries. OPEC countries are making public comment on the issue, and leaders of OPEC and the OPEC Fund have gone on record in recent times as saying that they are anxious to try to divert some of their surpluses to the Third World.

Of course, all of this requires institutional arrangements and infrastructures which are not yet in place. And I am reminded of what happened in 1974 when, in the Western world or in the Untied States of America, one of the OPEC countries bought a 25 percent interest in a major enterprise. This initially generated considerable fears when it was viewed as a portent of the future.

The moment a country with a large amount of money starts looking in your direction, whether you are an economic power like the United States of America or whether you are Jamaica, you begin to wonder if that country is going to buy the floor under your feet and the ceiling over your head.

Developing countries are the ones who have been preaching that we should try to be a master in our own house, and it took awhile for some of the Western countries to recognize that there was some virtue in the argument that we should, for instance, have sovereignty over our resources and not want everything to be owned by somebody else.

A 1974 editorial in the *New York Times* commented on this issue: "As Senator Metzenbaum of Ohio has pointed out, it would cost the oil states only 75 percent of their excess dollar earnings in a single year to acquire a controlling interest in eleven giant corporations, including AT&T, Boeing, General Motors, IBM, ITT, U.S. Steel, and Xerox." The piece argued that Congress should set maximum limits and full disclosure in advance of foreign acquisition of certain American industries, especially those critical to national security. But it concluded that "concerted Western policies for conservation and energy research and development are the fundamental ways of responding to the danger of excessive transfer of economic and political power to the oil states."

Interestingly enough, in the Federal Republic of Germany in December of the same year, a newspaper had this to say following the purchase by Kuwait of a block of shares in the Mercedes firm: "When oil-exporting countries produce their wallets to buy goods made in Germany, they are undermining the efficiency of our industry. When they start buying our industry and naturally enough concentrate on picking out the plums, they cause consternation and disapproval."

We have come a long way since 1974, and it's clear that the OPEC countries haven't bought out the Western countries lock, stock, and barrel, but it does represent an interesting phenomenon. And from the point of view of people like myself, those who provide the money do not necessarily have the right of ownership to everything in sight. Partnerships must form in the world between people who have the capital and those who have other resources, and the recent OPEC situation highlights this idea.

Finally, the question of the attempts to establish a South-South partnership. This has already been mentioned. It has been a very tricky road. The OPEC countries, feeling the pressures from all sides, from 1974 onward, have been very cautious in approaching any form of partnership with the West or even with the developing countries.

They have a political desire for a partnership with the developing countries on many North-South issues, but it has taken more political skill than we have produced so far to achieve this partnership. We have made, I think, substantial progress in the last year to the point where we could begin to talk about global negotiations based on the understanding that energy would be involved, but only if it were linked with issues of money and finance and a few others.

What we in the developing countries must learn is that you have to use all your political devices and skills if you are going to create that sort of partnership. So the fate of the global round of negotiations and of South-South cooperation will depend very much on our skill in trying to bring together the various elements within the Group of 77.

Chairman Weintraub: Our third speaker on this panel is John Foster, senior economic adviser of Petro Canada.

A View from the North

John Foster: I would like to provide a basic framework which would lead to subsequent discussion of the difficult issues. Let me selectively state some of the issues with which the wide range of industrialized countries are concerned.

During the last ten years, the energy issue clearly has become a critical one for the entire world community. Furthermore, energy is a complex and a controversial issue with a lot of money attached to it. There is a growing awareness that the world will soon have to make the transition from low-cost oil to much higher-cost sources of energy, including high-cost oil itself.

The availability of cheap energy in the North had been taken for granted by virtually all countries. Indeed, that is probably true of the Southern countries themselves, including the oil-exporters who were resigned to relatively low oil prices in the markets prior to 1973. Energy was not a focus of attention; it was forgotten by policymakers in most countries and was not a subject for the public at large.

Today, it is not possible to frame national economic and foreign policies without reference to energy, and this has meant that those concerned have had to master the key elements in the world energy situation.

We need to consider all forms of energy and not simply oil. Nevertheless, no other natural resource has turned out to be so strategic within nations and between nations as oil, hence the preoccupation with it.

It would be particularly helpful to search out the interests of the three major parties involved — the oil-exporting developing countries, the oil-importing developing countries, and the industrialized countries — and see where they conflict and where they converge.

First, energy is one of the most important imputs in the process of development, and consequently there is proper concern whether commercial energy will be available in adequate supply and at prices which will allow economic growth to continue. Moreover, traditional forms of energy are in jeopardy in the Third World, where they are the dominant and indispensable source of energy, particularly in rural areas.

There are probably about two dozen countries commonly classified as industrialized countries. Nearly all are members of the OECD and the International Energy Agency (IEA), the prime exception being France.

Like the countries in the other two groups, the industrialized countries are a very diverse group of nations, and their interests in energy differ significantly. Two of them, in fact, are themselves oil-exporting countries, Norway and the United Kingdom, and Canada is a large net exporter of natural gas in particular and, to a lesser extent, energy in general.

The industrialized countries' prime concern today must be energy security. Nearly all of them are dependent on imported oil for a significant part of their energy requirements, particularly for uses which cannot be easily replaced by other forms of energy, and this dependence will likely continue for the foreseeable future.

Hence, the industrialized countries are concerned with the economic implications of continued dependence on imported oil and also on the strategic significance of the fact that the major source of these imports is the Middle East, an area regarded by some of the industrialized countries and by some of the people in the industrialized countries, rightly or wrongly, as politically unstable and increasingly vulnerable to pressures from the Soviet Union.

In this context, there are probably six major concerns of the industrialized countries on which there is broad consensus with respect to energy. The first of these is that there should be an assured supply of oil from the oil-exporting developing countries. This means that the industrialized countries themselves have a stake in continued investment in exploration and development of oil resources in the major oil-exporting countries which, like Saudi Arabia, are the ones that still have the greatest potential for expansion.

The second shared concern is that oil for import should be available at what they consider to be "reasonable prices," clearly a tendentious phrase. Who can determine what a reasonable price is? The outlook for domestic energy prices should be such that their anticipated level justifies investment domestically in a broad range of alternatives. Hence, the industrialized countries themselves, by and large, have a vested interest in the new high cost of energy.

In addition, industrialized countries seek some predictability, over time, for the real price of oil imports. Assuming that the trend is still for an increase in the real prices of oil, oil-importing countries would probably settle for a gradual continuous increase and seek to avoid discontinuous jumps because these jumps are too much for the economic system to handle well. Their belief is that world output and their own output suffer severely and that the losses are probably much higher than they would be with gradual price changes.

A third shared concern is that additional improvements in energy efficiency must be sought. For the industrialized countries as a whole during the 1970s, energy used per unit of economic output significantly declined.

The industrialized countries would equally subscribe to the view that appropriate pricing policies would ensure further substantial improvements in energy efficiency to achieve the objective which those countries have set for themselves in various summit meetings. They are also concerned that the growth in energy consumption of this decade should average not more than 60 percent of the growth in economic output. This energy coefficient of 0.6 percent would represent a sharp reduction from historical ratios.

The next concern, financing of current account deficits, or the recycling of the OPEC current surpluses, remains a difficult problem. The present expectations are that the OPEC surplus, which was about $115 billion in 1980, will decline less rapidly than in the period after 1974 for a variety of reasons.

Moreover, the current account deficit of the industrialized countries themselves, which they're primarily concerned about, will decline steadily but slowly because of reduced economic growth, and they see the current account deficit of the oil-importing developing countries still there and increasing further before starting to decline.

The next point on which industrialized countries are agreed is that of cooperation among themselves on energy policy through such intergovernmental groups as the OECD, the IEA, and the European Economic Community. This includes steps to cooperate in reducing oil dependence through energy conservation, development of alternative sources, energy research, development, and demonstration, and plans for official stockpiling of oil and sharing of oil supplies in the event of an emergency. To this end, they have established various targets, usually set so they can easily meet them.

Now what does all this mean for policy options? There are several levels of concern: national, regional, and global.

At the national level, particular areas for concentration include those alluded to earlier: increased energy-efficient development, accelerated energy research, development, and demonstration, and the most fundamental one of restructuring their economies.

Let me say a brief word on energy efficiency. In the industrialized countries, as elsewhere, policies for energy conservation are an extremely crucial part of adjustment. The era of cheap oil prices has encouraged patterns of energy consumption and choices of technology which are simply inappropriate at today's prices, and so we're stuck with the baggage of the past. One of the problems in adjustment is that some consumption patterns are almost frozen, as in the case of private automobiles and public highway systems. Change requires political courage. Almost all industrialized countries are pledged to using the price mechanism as part of their energy policy, and such a policy is linked to a range of difficult domestic policies including income distribution and inflation. On the whole, I think much has been accomplished in the last two or three years in this field.

Some people have been thinking about whether some kind of global compact might be possible in the field of oil supplies and pricing. Let's start with the premise that this no longer looks like a promising line of approach. First and foremost, from the point of view of the OPEC countries, this is a nonnegotiable item, as we understand it. They have sovereignty over supply and price, and they don't intend to let go of it.

Nevertheless, there are still some worthwhile areas which industrialized countries might look at in the energy field to see whether the parties involved could discover a mutual interest. Areas of consideration include shared interests between the oil-exporting developing countries and the industrialized countries regarding the fate of the oil-importing developing countries in two respects: financing their balance of payments — the shorter-term problem — and financing their investment in energy supplies.

The third field is cooperation in research and development efforts, and here there are several options — perhaps more limited but very worthwhile — that can be explored on a regional basis of cooperation.

Then, there are questions of what kind of partial assurances can be given by way of supply and oil pricing.

Of course, all these things are a "can of worms," and I don't know if anybody has the answer to them. First of all, regarding the financing of balance of payments, the oil-importing developing countries face a two-fold energy problem: the short-term crisis and a long-term transition.

In the short term, they have to cope with the sharp increases in their oil import bills while somehow or other not losing the momentum behind their economic development. As we know, from 1974 to 1978, commercial banks carried the major burden of recycling the resources to the developing countries in deficit, but two-thirds of these credits were concentrated in only about ten middle-income developing countries. And now, these commercial banks are expressing concern about their portfolio limits.

Hence, in order to broaden the base of recycling of funds, it's inevitable that international financial institutions will have to play a greater role in the channeling of resources to the developing countries as well as in the direct investment process.

Now, regarding the financing of the energy investments of oil-importing developing countries, these needs are estimated by the World Bank at about $80 billion a year in current dollars during this next decade. The limelight, so far, has been very much on exploration and development of oil and gas, but the development of other energy alternatives is just as important, whether they be the new and renewable forms of energy or the commercial forms.

Furthermore, official development assistance is needed in almost all cases where domestic energy developments are unlikely to lead to exportable surplus. In the case of oil, this is fairly obvious. Foreign oil companies are not generally interested in exploring for oil in a country where they don't think the prospects will lead to exportable oil. They would say they are wasting their time. And when they say that the prospects for those countries are poor, what that means is that prospects for oil exports are poor. It doesn't mean there isn't prospect for small oil fields worthwhile to those nations.

The same argument can be applied to most development projects for other forms of energy. So this becomes part of the argument for official development assistance to overcome nonfinancial weaknesses in the field of policy advice and to strengthen planning, management, and training programs. In this regard, I would deeply regret any putting aside of the World Bank energy affiliate proposal.

There are also a number of bilateral assistance programs in this field, and new initiatives are being developed. I might mention by way of illustration that my own company, as a matter of government policy, is about to create a subsidiary which would do nothing but look for oil and gas in developing countries where multinational companies might be reluctant to go. This subsidiary would be financed out of official development assistance in the amount of Can$50 million each year. The total investment needs in energy in the developing countries are so large that there is room for all these initiatives and great merit in having diversification.

Now to mention cooperation in research and development efforts. By far the largest part of the world's research and development effort, outside of the Soviet bloc, is carried out in the OECD countries. The problem is to make this effort responsive to the needs of the developing countries and to bring them into the family. One of the major global issues is how to share the results of this Northern research with those Southern countries not yet in a position to contribute either manpower or capital for this effort.

Regional policy options or possibilities include (particularly in Latin America) joint arrangements for pipelines, research, and so forth.

Speaking from the point of view of the North, the final difficult area is that of supply assurances to the oil-importing industrialized countries. Reasonable supply assurances for oil imports, whether to industrialized countries or to the oil-importing developing countries, is not an area of divergence between the parties involved but one worth discussion. For example, supply disruptions may be either accidental or deliberate. In cases of accidental disruption, how are the oil supplies that go into the international trade to be shared? The IEA countries, for their part, have set up arrangements for strategic oil stockpiling and emergency sharing of supplies, but there is no such security arrangement for the oil-importing developing countries. This problem might make for some commonality that could be explored.

As for deliberate disruptions, such as the 1973 Arab oil embargo, I suggest that we dismiss wild thoughts of military intervention. But we need to ask what those oil-exporting developing countries who feel the need for the "oil weapon" want in return for giving those supply assurances.

Chairman Weintraub: Gerry Helleiner.

Panel Discussion

Gerald Helleiner: The question which John Foster ended with is implicit in both of the two previous presentations. I believe Jorge Navarrete made mention of a more predictable energy sector, and Donald Mills also referred to it several times.

It was stated that a global compact on oil pricing is now not terribly promising. The reason for it is the the OPEC countries believe they have gained control over oil prices.

It remains to be seen whether they actually have the capacity to control prices, and it will depend in particular on the power of the Saudis to police that kind of agreement. Nevertheless, they have devised a formula based on three indexes: indexes of exchange rate change, overall price change, and the rate of growth of the OECD, which they say they intend to introduce as the basis for future pricing of oil.

Let's take this at its face value and assume they can agree on this formula and abide by it. The question then is: What do they want and what does the North want? The North wants the oil to keep coming with steadiness in whatever increases in price the producers themselves have announced. And in order for that to happen, the producing countries, and particularly Saudi Arabia, have to be assured that the financial assets which they acquire through oil sales will earn a rate of return which is at least as good as that which they would have earned if they had kept the oil in the ground.

From the period between 1973–75 to the present, the rate of return they earned on their financial assets was negative or zero. The rate of global price inflation was greater than the interest rate they were earning on these pieces of paper which they acquired for pumping out their oil. They traded oil in the ground, which experienced rapid real price increases, for pieces of paper which were not paying them anything.

They have learned from that. There is always the possibility of wars and insurrections and other things, but let's assume they have the power to maintain their "formula" price. It follows that, in order to make it easy for the Saudi authorities to hold down the radicals on the domestic front, the North should offer a guaranteed real rate of interest which corresponds to that rate of increase in the real price of oil which these countries believe they are going to achieve. In real terms, this is probably not going to be more than 2 or 3 percent.

Would Saudi Arabia and the OPEC countries take such an offer with interest rates where they are now and with the prospect of the

Thatcher and Reagan administrations stopping inflation, whatever the cost? Unlike the situation of a couple of years ago, it begins to look as if the OPEC countries might earn, in fact, very high real rates of interest instead of negative ones if they *don't* accept an offer of this kind.

But it seems to me, nevertheless, they are inclined to be conservative. They are supposed to like to avoid taking risks. We should also avoid taking risks this big. It follows that the North should offer a decent guaranteed real rate of return on financial assets to correspond with the expected rate of real price increases to be enjoyed by the oil-exporting countries. Whether the expected price increases are actually realized is beside the point. The offer with respect to financial assets should still be made.

In fact, it seems to me we have the makings of a deal, and while that's going on, of course, you do all the other necessary things to protect the poor oil-importing countries. My first concern is still with them, but we are now talking about North-South bargains.

Chairman Weintraub: Paul Streeten.

Paul Streeten: I want to add a few points to Gerry Helleiner's scheme. But before I do that, perhaps I might sketch two possible scenarios, one which I regard as not very sensible and another which would be more sensible and in line with his.

The nonsensical scenario would be something like a continuation of what we have had in the past, assuming that we don't have in the next ten or fifteen years a large surplus of energy and oil in the world. A continuation of the past would mean very large and unpredictably timed increases in the price of oil, which in turn cause many other prices to go up because many products are made out of oil or oil products.

We have an additional impetus to inflationary price increases not only in oil but throughout the economy. A very popular doctrine claims that inflationary price increases have to be fought by monetary restrictions; that is, these oil-caused inflationary price increases are met by monetary restrictions which intentionally cause large-scale unemployment and significant underutilization of industrial capacity far greater than is required to reduce the demand for oil.

We don't actually reduce the demand for the oil very much through price increases; demand is not very responsive to these price increases. What is more, we actually cut off our noses to spite our faces because through deflation we deprive ourselves of the resources which could have gone into conservation and the exploration of alternatives. Therefore, the process repeats itself, and we get the next large increase. That is what I call the nonsensical, irrational scenario.

The sensible scenario seems to be something like this. It consists of six points, some in addition to Gerry Helleiner's. If we are assuming that for the next ten to fifteen years the real price of oil is going to increase, then we surely ought to agree on a gradual and systematic increase rather than on these upward and downward lurches in the real price. That's the first point.

The second point is that we should then agree on investment in oil exploration, oil substitution, and conservation, all of which would reduce our dependence and increase the supply of oil.

The third point is exactly Professor Helleiner's, namely, that there should be some form of guarantee of adequate returns on the financial surpluses earned by the four or five OPEC surplus countries. Not all of the OPEC countries are surplus countries; only some have substantial surpluses in their current account. And if we want them to supply a regular flow of oil to the rest of the world, then they must be given adequate incentives for the surpluses that they earn through supplying us with the oil we need.

The fourth point is that we should graft onto this scheme some mechanism that would channel if not the whole then at least some part of these financial surpluses toward developing countries. This is, after all, where there is the greatest need for finances, the greatest opportunities for raising standards of living, and the unutilized labor that can be productively employed with these financial surpluses.

The fifth feature of a rational scenario would be some kind of interest subsidy scheme which would guarantee that not only the middle-income countries — those which are growing faster and, in a sense, need the money least — but also the poorest countries would receive some significant proportion of these oil surpluses.

That means that we must build some kind of concessionary element through interest subsidies in some form or other for the benefit of the poorest countries, who need the most cushioning against these oil price increases and who need these financial resources most urgently.

The sixth element is that we must then institute a world trade regime that will allow countries that borrow these surpluses to service their debts through export earnings. You can only service international debts by exporting, and you can only export if the creditor countries allow for an increased demand for the debtors' exports. In other words, we must prevent the rise of protectionist barriers that would keep the developing countries that borrow a proportion of these surpluses from defaulting on that debt simply because there is no demand for their exports.

This is the rational scenario, and it is to my mind the most outstanding case of the so-called common or mutual interest thesis. The three groups of countries sharing this interest are the OPEC countries, the industrial countries, and the developing countries.

Why is there a clear mutual interest in having some kind of scheme like this? Because if there's one thing economists know, it's that normally you cannot have your cake and eat it. Doing one thing means doing without another.

But in the energy field, we have an outstanding example to the contrary. One item, oil, is in scarce supply and that prevents the utilization of capacity and the growth of world income. Breaking this bottleneck will allow fuller utilization of productive factors all around, so that if we invest in more energy, far from depriving other sectors of resources, we actually increase the degree of utilization of other sectors and we can all move ahead. It's an outstanding example of common interest.

Now, you might say that some vested interests are hurt, namely those who benefit greatly from a very high price of oil because increased supply may reduce the price somewhat below what it would otherwise have been. But even that, to my mind, is doubtful because the four or five OPEC surplus countries do not in fact now use their full monopoly power. They are keeping the price below what they could charge because they want to maintain political stability and do not wish to upset the Western economies too much. Sheik Yamani always says that it is in the interest of the OPEC surplus countries, such as Saudi Arabia, for the rest of the world to economize in the use of oil.

So there you have an example of an action which represents the common interest of the world as a whole, the industrial countries, the developing countries, and the OPEC countries. If these six points were adopted, we could create a rational world system of energy, finance, and trade. But if we follow the first scenario, we are simply impoverishing ourselves.

Chairman Weintraub: Joan Spero.

Joan Spero: What I hear around the table is great agreement about what we need to do. We all agree that we need to develop energy capabilities in developing countries, that we need to recycle, that we need to assure a more secure and predictable energy environment.

But the problem is how you do all of that. We all agree that we need to have recycling, and we recognize the need for greater involvement of international institutions in that recycling process. The question I would like to pose to the two representatives of the World Bank is this: Just what scope is there for that possibility in the next, say, five years?

I would like to take this idea one step further and make a few comments on what Gerry Helleiner said about a trade-off between, as I understand it, the OPEC long-term strategy and the guarantee of financial assets of the capital surplus oil-exporting countries.

The guarantee of financial assets was something that was discussed extensively at CIEC and nobody could figure out how to do it. It's linked with this question of recycling. When most of the recycling is carried out in the private sector and through Euro-markets, what mechanisms can you devise which would provide sufficient guarantees for a sufficient part of those assets and which would induce the capital surplus countries to enter into some kind of political bargain?

A footnote to all of this is that OPEC has not been asking about financial guarantees recently. Although we heard a lot about it at the UN, we did not hear a lot about it in other forums. I question whether that is feasible, and, if it's feasible, whether OPEC states are really all that troubled by it and whether it's enough of a political inducement for them.

Now, a question about the other half of the trade-off. Even assuming that a guarantee of financial assets would be enough to attract the oil-producers into this bargain, the question is whether they can pull off the long-term strategy themselves. The long-term strategy looked very good before the Iran-Iraq war, and the reason that it was never agreed to was partly because there were those who didn't want it. Certain "hard-line" OPEC countries were never interested in the long-term strategy.

Furthermore, I question whether the OPEC countries themselves are interested in the deal and, if they are, whether they can deliver in the long run. That is, if they commit themselves to some kind of stability of price and predictability of supply, leaving aside levels of price, will they be able, given the great possibility of revolution or war, to deliver agreed-upon supplies? And if they can't deliver on that — which seems likely — then what incentive is there for the developed countries to provide guarantees of financial assets?

Chairman Weintraub: Mahbub ul Haq.

Mahbub ul Haq: Since there has been so much consensus, let me strike a discordant note on this issue. I must say that I am not too enthusiastic about those who are searching for a global compact on the price and supply of oil, because I think the search for a global compact is based on certain very faulty assumptions.

First, the assumption that OPEC can control the price of oil is vastly exaggerated, because the price of oil depends on supply, demand, conservation measures, new investments undertaken in the next few years, and the price of substitutes. Out of all this array of factors, there is only one that OPEC controls, not the other four. And even there, present supply is far in excess of the needs of OPEC. Without OPEC the price of oil today would probably be higher, not lower.

There is also an assumption that the shocks which come from time to time and cause a sudden rise in the price of oil are disruptive. That contradicts the discussion we had earlier, which concluded that only a crisis creates the will for adjustment.

In 1973 and 1974, there was a major increase in the price of oil. Many countries took it for granted that the price would soon drop. An adjustment process didn't start. The period 1979–1980 saw another shock, and this time some countries became more serious about the adjustment problem. So I'm not quite sure that incremental increases create more will for change and for adjustment than shocks from time to time.

The supply agreements are now seen as vital for the survival of the North. Yet, we have lived for a long time in a world with all sorts of crucial commodity trade. In the case of food grains, 80 percent of the export trade is controlled by two countries. I don't know whether we are now going to seek supply arrangements in a whole range of commodities, or why we suddenly feel that the market mechanisms can't work.

I think the basic problems concerning energy are that we have gone from an artificially cheap era of energy to a high-cost era which reflects the true cost of energy. And by all projections, the price of energy is expected to rise by at least 3 percent per annum and perhaps more over the next two decades in real terms.

This creates tremendous adjustment problems in our consumption patterns, in our development patterns, and in our investment patterns. The necessary adjustments in organization patterns, transport systems, industrial structures, and technology are going to be traumatic for many countries, and no easy options are available.

So we must adjust development strategies to a new era of high-cost energy. Within that, we must seek energy efficiency, conservation, and other appropriate measures that reflect the correct price of energy.

We can soften some of the blow during the transitional period by having better recycling mechanisms. But the failure of recycling mechanisms today is due to a failure of the monetary system, not to the oil crisis. If we had a better monetary system, we could match surpluses and deficits.

In response to Joan Spero's question, my view is that if the international institutions are given the authority for effective inter-mediation in the market to recycle OPEC surpluses — for example, if the World Bank changes its gearing ratio, and if the IMF can borrow from the market and recycle — there are intermediation possibilities for matching surpluses and deficits.

And the issue of the guarantee of value of financial surpluses may be valuable to OPEC. This is one of its priorities. But there are various other ways of taking care of the financial surpluses of OPEC. It's not OPEC that creates financial surpluses; it's industrialized countries in their demand for oil.

As far as OPEC countries are concerned, they would much rather have lower surpluses because the economic rationale for taking more oil out of the ground than they need for their own development purposes each year is not proven yet. But if we are really serious about protecting financial surpluses, why not give OPEC an opportunity to invest them in real assets in developing countries and industrial countries? One of the OPEC functionaries was recently complaining to me that they had tried to buy a 10 percent share in an oil company in a Western country and were turned down. This was the same oil company which for several decades owned most of the oil production in that country. So who is sensitive about the issues of foreign investment, control, and participation?

Basically, the problems that have come with the so-called energy crisis are illustrative of the problems of the world crisis. The energy crisis has not led to the world crisis; rather, it is an outcome of the world crisis in a system which is already overloaded by pressures on nonrenewable resources.

As a result, we have to go back and look at the various elements of the world structure and try not to focus exclusively on energy. That's why people resist a separate global compact on energy, in addition to the fact that it is impossible for OPEC to give a guarantee on oil price because they cannot control it.

John Sewell: Mahbub's remarks were very interesting and offered an opportunity to find out where we disagree. Here the politics of the whole issue of global negotiations and relations between North and South enter in very clearly.

It's true that if one is to have a series of global agreements, energy cannot be isolated. But it seems equally true that if there is a lever on Northern decisionmakers, it's energy. It is, of course, quite right that energy is not the only commodity that one should be worried about. But it's the only one the North is worried about. We're not really worried about food because we export it rather than import it, as does Europe. So if you want a lever on Northern policymakers, energy and, in particular, oil become very crucial.

It's worth going back over what's happened. People in the North very slowly realized that we ought to come to some sort of agreement on energy, even if it wasn't a formal global compact, because it was in our interest to do so.

But the striking thing is that absolutely nothing happened. I think there are several reasons why — one of them being the perception of the Northern policymakers that the members of OPEC won't participate in this kind of effort. The North seems to feel that even if, for a number of reasons, OPEC did participate, an agreement wouldn't hold because OPEC is not cohesive.

And there is the notion that, even after you make your calculations, at least for the United States, it doesn't matter because, as long as you deal with Saudi Arabia and one or two other states, you have solved your problem in the short run. Therefore, trying to deal with the diverse membership of OPEC, let alone the other oil-exporting states, isn't worth it.

And finally, it's not politically possible in the North to guarantee loans or earnings in real terms. Returns for the oil producers' investments in the North may be quite logical from the economists' standpoint, but it's not very logical from the policymakers' standpoint, particularly in a country like ours which is busily cutting out all of the government guarantees of domestic loan programs because they run contrary to economic orthodoxy.

So in this area you've got a real dilemma, to the extent that energy either can't or won't be part of the bargain — and by energy, what we really mean, at least in the short run, is oil. And as long as it's not part of the bargain, implicitly or explicitly, the whole set of North-South discussions and the prospects for understanding become much less salient for any Northern policymaker in a political office.

And that, it seems to me, is one of the most crucial dilemmas. I happen to believe that if the oil weapon had been used in a global sense in the interest of international economic reform, rather than for particular geographical, political interest, it would have had a marvelously stimulating, mind-clearing effect on decisionmaking relating to a broad range of other international economic issues. But it hasn't.

To speak quite frankly, all of the signals that have come out of bilateral discussions are precisely the opposite. It's the outsiders' perception, for instance, that one of the elements blocking agreement in the UN in the last round was the fact that, either explicitly or implicitly, the surplus oil producers and most of the industrial countries weren't at all interested in discussing financial issues within the UN forum.

Now, if this is the case, the prospects for North-South negotiations are miserable indeed. And I think one ought to be quite frank about that from the standpoint of Northern policymakers, despite that fact that I agree completely with the prospects, in the theoretical sense, for the kind of positive sum game on oil issues that Paul and Gerry and others have outlined.

Chairman Weintraub: Nicolas Barletta.

Nicolas Barletta: The discussion illustrates that there may be agreement as to where the problems are, but not how to go about solving them. Part of this lack of agreement is related to the fact that we are not being precise enough as to the nature of the problem.

In the energy field many of the problems could be synthesized perhaps into two areas: the long-run supply problem at the most reasonable price or at the lowest rate of price increase given the scarcity worldwide, and the short-run adjustment that everybody has to make to the very wide fluctuations in prices.

The short-term adjustments are obviously creating problems for everybody, but especially for the developing countries that have a lesser capacity to cope with them. The massive transfer of income that these countries are having to pay, especially the ones that import oil, can only be taken care of either by increased debt or by arresting growth and consumption, generating more unemployment, and imposing sacrifices on their population.

Given the tremendous challenges the developing countries already face in promoting growth and development, this short-run problem has to be met immediately, and I frankly don't see how any global negotiations can cope with that problem.

The developing countries are having to slow down growth. They're having to make immediate adjustment to the new situation in their balance of payments or make medium-term adjustments internally in their economies, which they are financing with additional debt. Most of the debt is incurred through recycling, but recycling is taking place through the private banking system by and large, even though a small part is taking place through the international and regional institutions.

So if we examine which mechanisms are working, the mechanism of negotiations or the mechanism of the institutions that are already in place, I frankly don't think that the short-term problem can be successfully addressed in global negotiations.

The crux of the recycling problem, which creates a disadvantage for the developing countries, is that the debt they are acquiring in order to succeed in the short-term and medium-term adjustment should be as long-term as possible. But the private banking system has not been able to provide such long-term financing.

This crucial issue is the one that could be addressed perhaps by some of the existing institutions if we were to create the energy affiliate that the World Bank is proposing or if the IMF manages to raise additional resources. The questions that have been raised about the credit-worthiness of the countries in the medium term also have to be answered.

The central point is that the short-run problems being created by the energy crisis can certainly not be addressed very well, as the evidence demonstrates, by global negotiations.

Thus the problems of short-term adjustment and medium-term restructuring of economies need to be taken care of anyway. The question is whether the price of that adjustment to the developing countries has been unnecessarily high. While the talk goes on, action has to take place, because the problem cannot wait.

However, the long-run supply problem could perhaps better be addressed in a form of global negotiations. One approach already defined here by some of my colleagues involves an exchange of supply guarantees for asset value guarantees. But a more realistic approach to that problem would be to produce more energy and conserve the energy that is available.

The aspect of production again comes back to financing. As Mahbub ul Haq has noted, there is great expense in just maintaining the present situation and keeping it from getting worse. It is going to require $80 billion of investment per year for this next decade to develop new sources of energy in the developing countries.

If we manage to address that issue, the increase in the production of energy in the developing countries, then I think the developed-country problem will tend to resolve itself. As Professor Rostow has mentioned, if the developed world, which has far more financial and technological resources, addresses realistically the issue of increasing domestic energy production and thereby makes the available supply more accessible for everybody else, then these two measures would be more realistic than any global negotiations on regulations and controls in helping to solve the medium-term problem.

Chairman Weintraub: Ed Hamilton.

Ed Hamilton: I agree with most of that and a good deal of what Mahbub and John have said. The interesting fact that has to be explained is the one that John Sewell raises: If there has been such consensus and the issue is so pressing, why has nothing much occurred?

The basic reason is that none of the major participants really wanted anything to occur. What they wanted was to maintain the sense of urgency which would keep everyone aware that, unless reasonable restraint was shown on all fronts and unless the adjustment process was made possible, something much more painful by comparison would occur.

The straightforward approach guaranteeing financial return and establishing reasonable ceilings and floors on the value of the paper that the OPEC countries are receiving seems to have been fairly clear to most people who look at the problem for any length of time. It's a more or less classic commodity adjustment problem.

The difficulty is that trying to apply this solution on the scale involved is so novel and even potentially dangerous that no one really wants to take this option unless all others have disappeared.

What has been proven since 1973 is that the OPEC countries, despite problems of internal cohesion and long-term stability, are extremely conservative and feel responsible for the world system and for maintaining it in some workable order. None of us, I suppose, could have expected quite as careful a management capacity as they have demonstrated, despite the fact that they don't strike one immediately as the most stable countries imaginable.

Of course, a lot of the feeling within OPEC against "sensible" pricing policies is precisely due to the instability among them, which also makes it possible somehow to manage the situation. That is to say, if OPEC were able to become monolithic and operate as a cartel without internal stresses of any great dimension, I doubt that the outcome would be better for either the developed or the less developed countries.

So surely we have to accept the premise that the felt need to formalize this *de facto* situation is not there. There seems to be less felt need now than there was four years ago when people were really scared. Furthermore, we've got to understand that the new American administration is likely to be very much more concerned with achieving energy self-sufficiency or at least energy security in the developed countries, and more specifically in North America, than in looking after the energy needs of any broader group of countries.

I would not be at all surprised if the new American administration started again talking about Mexico, Canada, and the United States as essentially an energy self-sufficient area. The talk would immediately be rebuffed by both Mexico and Canada, but it seems possible that the idea might gain some currency and maybe eventually become a basis for negotiation. It might well be a broader concept and involve some of the other countries in Latin America.

And the odds are good that agreement on energy is likely at least in formal terms. Whether or not it becomes operational is another matter. But energy, in formal terms, may well become the first experimental proving ground for the regionalism that Walt Rostow has spoken about.

You may well see attempts to design various regional approaches to energy, but I think the main effort in that kind of negotiation will come from the very large producers on the one hand and the developed country consumers on the other. The issue will be very much as it is with special drawing rights (SDRs) — how to get the development problem hitched to something which is essentially going on between two groups who are dealing in terms of their own needs and insecurities as distinguished from the development problem directly.

I don't mean by that to denigrate the attempt but to point out that this situation makes it harder, not easier, and it therefore requires more innovative thought.

Furthermore, if we really wanted to find a way to handle the problem in the ways that Gerry and Paul are talking about, I don't think it's really a very difficult technical problem. That is to say, we are experts at finding technical solutions. Our entire economies are built on mechanisms which balance ownership with control without disrupting the whole system.

I can imagine the Export-Import Bank selling participation, for example, to the OPEC countries with all kinds of fancy rules about the voting rights. This might well satisfy OPEC and at the same time give it real assets as distinguished from paper. We've done that in thousands and thousands of variations. We have invented all kinds of financial instruments designed essentially to manage that problem, and this just doesn't seem to be any greater challenge than those.

My own hunch is that the volume of research and development which has been launched on the energy problem is being underestimated fairly systematically in the discussion. I would not be at all surprised if we had some very interesting technological breakthroughs on the energy front in the medium term. I think the next ten to fifteen years will produce some very interesting developments in synthetic fuels and various other kinds of energy innovations.

Walt Rostow was arguing that development of domestic energy sources could produce self-sufficiency in the United States by 1990. That would strike some of our energy experts as a hair too optimistic; nevertheless, a substantial technological breakthrough seems to me to be in the wind. Of course, that's both good and bad news for the less developed countries. It's perfectly true that it will leave a lot more oil for other countries, but it will also very likely change the economics of world production.

It may well be that, whatever the technology or set of technologies that makes oil obsolete, it will then become very hard for an oil-based economy to compete in world markets.

Chairman Weintraub: I am inspired to make two comments.

Not too long after 1974, we devised a scheme whereby we sold Roosa-type bonds to Saudi Arabia and guaranteed them. The program died for many reasons, but it shows that we can devise solutions if we have to. I don't really think that's the issue.

The second point I want to make concerns the research and development going on now as well as the claim that little adjustment has occurred. People say that the shock didn't really matter, but it's just not so. The energy efficiency ratio in the United States is now well below what it measured in the early 1970s. Price does make a rather substantial difference on our consumption patterns.

All of you are assuming that real prices will increase 3 percent per year over the next two decades and that there's going to be an oil-short economy; therefore, we have to guarantee OPEC financial assets in some way in order to assure ourselves of oil supplies.

This could be the right conclusion, but it could also be the wrong projection of the demand for oil. There's a respectable body of opinion that thinks the oil shortage projection is incorrect.

Roger Hansen: In a conference addressing North-South problems in the 1980s, doesn't some of the evidence presented in this session on energy raise a very fundamental question, namely, will the phrases "North-South conflict" and "North-South accommodations" have any empirical meaning in the 1980s? Indeed, may the time come when those people who for fifteen years have predicted that the South would disintegrate as a diplomatic bargaining unit and that, therefore, the South's demands were of little concern, are finally being proved to have the more valid argument?

The reason I suggest that we all ought to consider this proposition relates directly to our energy discussion. If energy, particularly oil, is not to be part of the Global Negotiation, then, like John Sewell, I wonder what is left to hold the South together as a diplomatic bargaining unit that can gain and hold the attention of the North. That is why Mahbub's pessimistic remarks about tying the energy issue to any global negotiations leave me wondering if we shouldn't all go home.

If one is prepared to be mildly pessimistic — and one tends to be after listening to Joan Spero's analysis of how this issue is never salient to the North — then the problem for the South has always been how to force the North to give it enough attention to permit preparation for serious negotiations.

If oil, or energy, is lost to the negotiation because the relationship among members of the Group of 77 — that is, the nonoil countries on the one hand and the oil countries on the other — is such that this issue cannot become a central element in an overall negotiation, hasn't the South lost the one arrow in its quiver? The energy issue alone caused enough concern in Northern governments and bureaucracies that the North was prepared, however reluctantly, to sit through seven and a half years of discussions at the United Nations and elsewhere. Indeed, the North is still prepared to travel to Cancún to continue talking even if it is not considering serious negotiations. If oil is dismissed, the saliency of North-South issues is lowered to the level of GS-9s in most Northern countries, if it is that high at present, and that will end even the empty dialogue of the past seven years.

Several other propositions follow from this, unfortunately. The North may turn more easily to a genteel neomercantilism in response to the difficult economic problems of the 1980s and in Trilateral

Commission fashion spin out agreements of the kind now being made between Japan and the United States over automobiles.

With just a slight bit of attention, the newly industrialized countries — whose capacity to export to Northern markets will also then have to be controlled — present no insurmountable problems to this kind of genteel neomercantilism. The number of countries is so small and their desire for access to the markets is so great that this kind of a negotiation in a back room will prove easy enough for people sharing a view that the system is working to handle.

If that's the case, then, what happens to the North-South negotiations until the set of "global" issues concerning resource scarcities, the environment, and food production become so salient that, like oil, they bring the North to the negotiating table once again?

This may be far too pessimistic. But if the North is as little interested in serious North-South negotiations as the experts and the evidence suggest and if the oil issue is not to be a major part of the negotiating process because the oil-producing and oil-consuming developing countries can't reach the necessary degree of agreement within the Group of 77, the North-South negotiation may truly have reached the point of a charade and perhaps we ought to lay it to rest.

Mahbub ul Haq: Mr. Chairman, I have a two-handed intervention to make. I want to make one thing clear: it's not that the energy issue is not negotiable or that it should not be part of global negotiations. Nobody is suggesting that.

In my judgment, there are certain parts of the energy issue which are negotiable and certain parts which are not. The parts which are negotiable concern production of energy and investment resources whether for energy-affiliated projects or any others. Similarly, the pooling of energy research at the regional and even the global level as well as the recycling mechanisms to take care of the short-term balance-of-payments problem and the medium-term adjustment are all part of the negotiable package.

What is not negotiable is the global price and supply agreement on energy because I don't see any reasonable way that it can be brought about. But we are not taking oil or energy off the global agenda. Many of us have suggested very definitely that it should be part of it, but there are only certain parts of it which can be negotiated.

Roger Hansen: Mr. Chairman, may I?

Chairman Weintraub: Roger.

Roger Hansen: I do apologize but it does strike me as a crucial issue.

My question to Mahbub would be whether or not the parts of the energy issue which he still feels are negotiable and should be part of a negotiation are important enough in the minds of Northern policy-makers to bring them to the table.

In other words, if indeed there is subtle agreement between the oil-exporting developing countries and the industrialized countries and if, to paraphrase Ed Hamilton, they share a significant range of interests and they're playing a game without the need for negotiation in any formal sense, then are the parts of the oil issue you mentioned serious enough to bring OPEC countries and the Northern states into a much wider negotiation? If they're not, shouldn't we simply adjourn and toast the memory of the North-South negotiations?

Chairman Weintraub: Joan Spero.

Joan Spero: I feel obliged to respond to Roger's question. Let me tell you the way I saw it, in terms of the different actors.

I agree with you that the Group of 77 felt that oil, not energy, is the one carrot. It's the inducement to talk about raw materials, trade, development, and money and finance. Now, there were a couple of problems with that.

One problem concerns the distinction between a global energy compact on the one hand and an ongoing energy forum on the other. For the Europeans and for the Japanese, the carrot was an energy forum. They very much wanted a place where you could talk about energy.

It's what the French wanted out of the CIEC back in 1974. They don't expect a compact, and they are interested in energy production for developing countries and the associated issues we have talked about. But the carrot for them is the forum — the institution where they can discuss energy issues with the oil producers.

So this is indeed enough to attract the Europeans and the Japanese. It was not particularly attractive for the United States for a variety of reasons. For one thing, we didn't think that much was achievable in an energy forum. We felt that it was important to talk about energy in terms of the developing countries, but we didn't think that we had much to gain from an energy dialogue.

Part of that attitude can be traced back to the different energy-dependent situations between the Europeans and Japanese on the one hand and the Americans on the other. Another dimension of relative United States indifference to the forum concept had to do with the credibility of the carrot.

What we constantly heard in the negotiations from the Group of 77 was that they would produce this energy dialogue, and yet the one thing we heard from the OPEC countries was that they did not want a separate *ad hoc* group in which energy would be negotiated.

Additionally, the OPEC countries constantly linked energy with monetary reform in the negotiations. That was the negotiating linkage that was of primary concern to them.

If you want to talk about energy, you will have to talk about the SDR-aid link and you will have to talk about decisionmaking in the IMF among other things. This has decreased the interest of the United States in negotiating energy in the Global Negotiation and also the interest of the European Community.

One can ask whether OPEC was really concerned with this linkage. Was that linkage made because OPEC was seriously concerned about monetary reforms, about decisionmaking in the IMF, and about the SDR-aid link, or was that link made because OPEC wanted to up the ante to a point where it would oblige the developed countries to say no to a negotiation OPEC never desired?

To summarize, energy was the carrot which the South tried to use to negotiate the North-South dialogue, but did the Group of 77 have that carrot to offer? And how interested was OPEC in being part of the dialogue? My suspicion is that OPEC — and I leave Mexico out of this, because Mexico takes a very different view about how to organize an energy dialogue — was never willing to play the role of the carrot.

Chairman Weintraub: Nick Barletta.

Nicolas Barletta: I agree with the perception expressed by Mahbub ul Haq of what should be negotiable in the energy field and what doesn't seem to be realistically negotiable.

The other important element would be the forum and the people in that forum, that is, the people in the political arena but with economic and technical expertise to deal with these issues, both in their short-term and medium-term aspects. And I wouldn't leave out the possibility of building blocks that could start at regional levels and then become globalized afterward as ways of implementing various solutions to the problem.

I would stress that the problem cannot wait for a very comprehensive, complex, and frustrating negotiation that may never get anywhere. The problem is already there and needs to be taken care of.

Chairman Weintraub: Jorge Navarrete.

Jorge Navarrete: The discussion has raised a lot of very interesting issues for comment. I will choose some of them which seem to me particularly interesting.

There was an initial suggestion that one of the trade-offs in negotiations on energy would involve stability or predictability of prices and supplies and security of financial assets. Maybe this kind of trade-off is interesting for some of the oil-exporting countries, but by no means is it interesting for all of them.

I suppose that many oil-exporting countries, some of those in OPEC and of course Mexico, are interested in another kind of trade-off. The range of objectives that these countries are seeking in bilateral dealings is much wider. It encompasses enhanced development opportunities in the areas of trade, industrial cooperation, technological advancement, financial assistance, and so forth.

We have also heard Professor Streeten present the idea of two scenarios, one nonsensical and another which may be not practical but is sensible. I would say that the nonsensical scenario is not limited to the energy field. In the world economy as a whole, we are in a nonsensical scenario, and if past tendencies not only in the energy field but in the world economy as a whole continue, it seems also that we will be unable to move out of the nonsensical scenario.

Joan Spero mentioned several times the question of whether the OPEC countries are reliable partners to negotiate with. I would stress that, if we adopt this point of view, the reliability of Northern states in international relations would appear equally questionable. For instance, a few months ago, none of the people involved in the Law of the Sea Conference had any idea that one of the major participants would suddenly cast all the previous negotiations in doubt and would want to review all the previous commitments, thereby calling into question the credibility of the entire negotiating process.

I am particularly interested in the interchange that took place between Roger Hansen and Mahbub ul Haq on what can be negotiated in the energy field and whether this is important enough to attract the serious attention of all parts of the international community, particularly the developed countries.

This interchange seems to be the best proof that in reality countries approach the problems of international negotiation from a very pragmatic and self-interested point of view. That's the reason why I stressed earlier that one of the bases for all these exercises in relation to the New International Economic Order, or the reorganization of an international economy, is precisely the possibility of finding common interests among the countries.

Maybe the negotiable package on energy that the developing countries can offer would not be attractive enough, and this is of course a risk for the future of the dialogue. It may be that the price they are asking for discussing energy in multilateral, intergovernmental organizations is also too high, as has been suggested here, but this is also part of the negotiating process.

A question came from the audience concerning my proposition that the energy questions must be linked with the others. For example, international discussions on supply of energy, to be feasible, need to be coordinated with discussions in other sectors of energy supply, such as the supply of technology, materials, and equipment for nuclear energy.

I am not suggesting a direct link between the two elements, but rather a wider approach to energy. If the interest of the oil-importing countries on predictability of supply is to be understood, the interest of other countries in predictability of supply of other energy sources also needs to be understood.

Chairman Weintraub: Donald Mills.

Donald Mills: I would like to comment on the question of whether the OPEC countries really want to discuss energy. One has to recognize that the Western countries have shown every desire to separate energy from everything else.

And we should not underestimate the feelings of the OPEC countries about the Paris conference. They insisted on extending the participation and the subject area to insure that energy was discussed in the context of other major development issues. But the effort to arrive at a broad compact in Paris failed.

We have been trying in the UN for some time to launch what is called the Global Round of negotiations on the NIEO. The Group of 77 at first proposed a central UN forum with authority to negotiate on all issues *ab initio*. But in the face of strong objections to this, a compromise proposal was worked out in which the central forum or conference would first review the various issues and then pass them on to various existing UN and other international institutions — for negotiation and for final consideration in the central body.

But the one subject for which there is as yet no institution for such negotiation is energy. Now, obviously, in the face of the well-known desire of the Western countries to take energy by itself and the well-known attitude of some Western countries of not yielding on monetary issues, it would be unacceptable to allow energy to be negotiated by itself in the central body because there was no forum. In any event, we are still short of full consensus in the UN on the mechanisms for the Global Round.

The one thing that describes this is a term that certainly came up privately more than once during all of these negotiations about a global round late last year: lack of trust. It's the major feature in the negotiating process. It's not a matter of distrust in the simplistic sense. It's lack of understanding and an insecurity about the people with whom you're dealing on both sides. It is a very important issue in international relations, and it has surfaced again in the discussions about whether OPEC countries in the future can carry out any guarantees they make in matters like pricing.

I would like to move on to the question raised by Roger Hansen: Will the South disintegrate as a political formation in terms of negotiation?

I would say that it could well happen. The negotiating process that we initiated seven or more years ago might disintegrate for a time, but that should offer no consolation or joy to any of us, because the issues are not ones which will go away.

The fact that we raised these issues in certain circumstances and expressed them in certain ways is really incidental. But the issues themselves are fundamental. There are, in my view, historical, political, and cultural circumstances which make demands of us as members of the world community. Paul has asked whether there is, in fact, a world community. Well, we find that we are living together and must come to grips with the world's greatest disparities, without necessarily attributing blame.

If you look to the year 1990 or the year 2000, it may well be that the North-South dialogue won't get very far. But if you look to 2020 or 2050, you realize that the imperative to reconcile interests is very great indeed. Much depends on whether we can rescue our present efforts, which are in serious trouble.

Finally, let me mention one sort of device which has been attempted in order to bring about partnerships in the Third World involving energy. Mexico and Jamaica have been trying to establish a relationship between Jamaica's bauxite and Mexico's oil. It is illustrative of the sort of partnership that could exist not only between South and South but between South and North.

Since Jamaica has no oil as yet but has bauxite, it would establish with Mexico a joint company in Jamaica for mining bauxite. It would be an additional mining operation jointly owned by the two countries, with Jamaica having the majority ownership and Mexico the minority. The product would be shipped to Mexico, where it would be processed into metal by a company jointly owned, with Mexico having the majority of shares and Jamaica the minority. The point is that if we are going to come together in some sort of way, we have to fashion new devices and new partnerships.

Chairman Weintraub: John Foster.

John Foster: I would like to respond to Ed Hamilton's remarks about North American self-sufficiency. Of course, this is a kind of myth in its own right. Both Canada and, I believe, Mexico are energy self-sufficient already. So what you mean by "North American self-sufficiency" is a lack of it in the United States.

What we have north of the border is exports on net balance in gas, electricity, and coal and net imports of oil, but the net export on overall balance is positive for energy.

So the real problem for Canada is the oil imbalance for the East Coast, for which supplies come from Mexico, Venezuela, and Saudi Arabia in particular. And there, we have to remember the long-term prospect of discoveries in the Atlantic — Hibernian oil and Sable Island gas — both of which look promising.

I will go on to my two major points. First of all, the energy issue is not just a matter of oil; there is also a fuel wood crisis which none of us mentions enough. Maybe we just assume that we all know about it and feel that oil is a preoccupation we really want to debate because it's very difficult. Also, we live with it now, and certainly in the industrialized countries it's what we're really bothered about.

Nevertheless, given the present rates of deforestation in the countries of the Sahel, Senegal will be bare of trees in thirty years, Ethiopia in twenty, and Burundi in seven. So a sizable percentage of the world's forest could disappear before the end of the century. For many countries, this is a matter of far greater concern than the short-term oil crisis.

Now for my final point. In my introductory comments, I steered clear of the long-term world energy prospects as a subject which is enormous in its own right.

It is extraordinary how quickly we discard the conventional wisdom of last year as being utter garbage and think that today we know the truth. But next year we will have quietly dismissed this year's garbage and blamed somebody else for having invented it.

This makes me feel, therefore, that the only worthwhile approach from which we may then start to prudently plan is the scenario approach, because we just do not know what the future is going to be.

For example, a French approach for a five-year plan for energy uses this technique. It has three scenarios. It has a "rosy" plan, a "gray" plan, and a "black" plan. I think we have probably been discussing the "gray" plan, and that's the one adopted there; the plan assumes a 7 percent per annum increase in real oil price internationally.

In the "black "plan, everything goes wrong you can conceive of, and that's the world of those people who specialize in the geopolitics of energy. It's an awful place, and yet you still need a plan for it.

The third one is the "rosy" plan, much like the one described by Professor Singer of the University of Virginia in an article showing all the wonderful things that can happen in terms of energy demand restraint, accelerated energy supply, and restructuring of economies, when all the sensible policy decisions are taken. On that basis, clearly, OPEC just goes away.

The Shell International view on this problem is one I quite like. It basically stresses two scenarios: restructured growth and the World in International Contradictions (WIC), within which their planners have to plan.

This brings to mind the price scenarios that come out of all this. I'm one of those who has, in the past, pushed the thesis that on present evidence real oil prices would seem to stay roughly where they are.

Politically, this has also been a convenient hypothesis. However, this clearly won't do after the second oil shock. So now a common conventional wisdom is that real oil price increases will be 2 or 3 percent (e.g., World Bank). Another is 7 percent per annum. Two highly esteemed United States consultants use those figures. Other people say that the real oil price might be about to collapse, and they point to 1982 as a year in which there might be a world oil glut and Saudi Arabia might have to curtail its output to three or four million barrels a day.

My feeling, confirmed by the French approach, is that, in terms of prudent investment planning for an oil-importing and oil-reliant nation, it is wise to assume the "gray" view, or better still, work within the broad range of the two scenarios. Perhaps it's cheating to divide by two and take the mean. But it's very dangerous to plan on the "rosy" scenario; we would be in real trouble if it didn't happen, and no nation that imports oil could afford to take that risk.

We just don't know where prices will go. They could head either way. The only thing that's certain is that we're going to be wrong in our predictions. This, I think, is what bothers me about global oil compacts: we might find ourselves planning for last year's war.

For example, real oil prices might explode, or they might suddenly break. I don't say that either of these things is likely to happen, but you can't be sure. Suppose oil exporters and importers did agree among themselves on a nice OPEC long-term formula of real oil price increase amounting to 2 or 3 percent. What would happen if the Saudis lost control of this because real oil demand suddenly bounced back once we moved out of recession and into excellent economic growth, thanks to good government decisions?

Or what happens if, on the other hand, through good government decisions and corporate planning and conservation measures, the oil demand disappears and the real oil price crashes? I don't say it's likely, but what would then happen to that price agreement, which requires OPEC countries to maintain a formula price path?

Are we just deluding ourselves when we use neat straight-line price increases because we rather like a tidy solution, even though it may not work at all?

Chairman Weintraub: I will just comment on what you have said about the uncertainty of the future market, and then I'll try to summarize.

I remember when oil prices started to go up. When oil was about $10 per barrel, everybody said that when the price reached $15, various kinds of synthetics, coal gasification, shale, and other alternatives would become feasible economic ventures. Then, when the price hit $15, it was felt that an increase to $20 would make the alternatives feasible ventures. No matter how high the price went, it was always some future price at which these things would be feasible.

Nevertheless, like other people who have been observing the energy situation, I have been impressed by what's been happening to the marginal energy use–output ratio. It's been going down quite dramatically. This fact has been left out of a lot of the discussions here, although it's really quite significant. It's one of the facts on which a good many of those "rosy" scenarios are built.

Let me see if I can summarize very briefly with two or three points.

Donald Mills has talked about the lack of trust in the negotiations, not whether people are telling the truth but whether people can do what they say they're going to do or whether they mean what they say in quite the way they say it.

I suspect there's a lot to that, and I'm struck by his reference to an Overseas Development Council report on new and renewable sources of energy and the idea that people might be afraid that developed countries would take the oil and leave everybody else the rest.

This had never occurred to me, and I rather doubt it has ever occurred to any other responsible American.

Donald Mills: I was also surprised. It's just a thought.

Chairman Weintraub: But you also cited another idea that bothered me on the question of nuclear proliferation: the concern that countries have to be able to get secure supplies of nuclear material if they're going to diversify their sources of energy. That's absolutely right, but at the same time, it's really not very fair to slight the real concern that existed about nuclear proliferation — the threat of world conflagration. There are irresponsible elements in the world as well as responsible elements.

The only reason I mention these concerns is that we — each of us — just don't believe what people say. It came up earlier with the idea of the basic human approach to development.

I remember when this subject was first broached. It was distrusted very deeply by the developing countries. It was perceived as a way to stop transferring real resources in any real amount and to keep them "hewers of wood and drawers of water."

This distrust is a problem that will permeate our discussions — not just in energy, where the distrust is great — but in other areas as well.

We have heard various reasons why a compact has not been possible: because OPEC is not interested, because the developed countries have not been interested, because the Group of 77 is not interested unless other things are attached to it, and so on. If nobody is interested, why does the idea keep recurring? There must be some interest somewhere.

I think we do agree that, in certain areas such as research and development and investment and recycling, there is scope for negotiation and, hopefully, some kind of agreement.

There are two views, a view from the North and a view from the South, concerning the interests of different groups of countries in energy. I was struck that both Mahbub ul Haq and Nicolas Barletta concluded on very strong national-interest grounds. I pick it up because I guess it's one of the starting points at which I tend to look at things.

Señor Navarrete explained that Mexico is not really interested in an agreement on financial security in return for the security of supply and price. Mexico is much more interested in using the leverage of oil bilaterally in order to diversify its trade and acquire other financial resources.

Therefore, this is not really a Southern view but a Mexican view, although the Mexican view may coincide with the Southern view. I'm not criticizing this stance because it strikes me as quite natural. But it's a national view.

I raise this point because it comes up over and over again. There are areas where interests coincide, but there are a good many areas where the views from the North and the South obviously diverge because each group has its own interests.

John Foster's comments reminded us of the uncertainty about future supply and price. And I want to recommend that all of us be a little skeptical of what we as "experts" say each year. Because we're probably wrong each year, too.

Session III
The Bretton Woods Institutions:
What Room for Accommodation?

Chairman Streeten: The two Bretton Woods institutions were created in 1944 and started to work in 1946. Located in Washington, D.C., they are the World Bank and the International Monetary Fund, or "the Bank" and "the Fund."

I am also going to discuss a third institution which was created later and which is not strictly a Bretton Woods institution. This is the General Agreement on Tariffs and Trade (GATT), which is concerned with trade issues.

The difference between what we might call the Bretton Woods institutions and the San Francisco-born institution, namely the UN and its specialized agencies, has perhaps been best characterized by Mahbub ul Haq. In his usual succinct, lucid, and witty way, he has suggested that this difference lies in the fact that the Bretton Woods institutions, the Bank and the Fund, have money. They also have, I might add, a certain amount of professional efficiency in the light of their own objectives, but they are not democratic because the voting is by the shares of money that countries contribute.

The UN and its specialized agencies, on the other hand, are democratic in that each country has one vote. Some people, perhaps, might criticize this system, because even quite small countries, with few people and very little money, have the same vote as the larger countries with many people and much money. What one would want from a reformed and ideal system is some form of democracy combined with the money to do the things we want to do efficiently.

Now, the main distinction between the two Bretton Woods institutions is that the International Monetary Fund lends for very short periods, and essentially for balance-of-payments support. If a country has problems because it has inadequate reserves and its imports exceed its exports, it can go to the Fund and, subject to increasing conditionality (that is, the conditions imposed by the Fund), can borrow money for short periods.

The Bank, on the other hand, lends for much longer periods for development purposes. It also has its conditions, but they are quite different from the conditions of the Fund. And the bulk of its lending is for projects like utilities, dams, roads, and, increasingly in the last ten or fifteen years, for projects in agriculture, education, nutrition, and most recently even in health.

The names of these institutions are somewhat misleading. It was Keynes who said that we normally call institutions that lend on short term "banks," and institutions that lend on long term "funds." Therefore, the Fund is really a bank, and the Bank is really a fund. The main thing to bear in mind is that the Fund lends on short term for stability, and the Bank lends on long term for development and growth.

The General Agreement on Tariffs and Trade regulates trade. It took the place of a third institution which was very much discussed in the 1940s but which never received United States support and never came off — the International Trade Organization. In many ways this organization and the concepts in forming it were very interesting and full of foresight, and it covered a much wider area than the General Agreement on Tariffs and Trade.

The GATT leaves out many areas that the people who negotiated the International Trade charter at Havana had in mind concerning the International Trade Organization: such as trade in agricultural products; trade between states or state organizations, which has become increasingly important; restrictive business practices; and trade between the affiliates, the subsidiaries, and the branches of multinational corporations, what is called "intrafirm trade." None of these areas are dealt with by the GATT. It is mainly about tariffs and nontariff barriers to trade in manufactured products.

The sequence of presentations will be as it is on the program. Mahbub ul Haq will begin with an overview. I will then say something about finance and development, and Gerald Helleiner will talk about trade. General discussion will then follow.

Overview: Altering Tasks, Changing Institutions

Mahbub ul Haq: Thank you, Paul.

Let me say, first of all, that I speak this morning in my personal capacity, a caveat which I think is important if I am to speak frankly and yet stay on with the World Bank.

I shall also focus a bit more on the World Bank in my presentation than on its sister organization, the International Monetary Fund, though I may have a few comments about the IMF.

Now, within the World Bank these days we have the usual good news and bad news about the future of Bretton Woods. The good news is the existence of a certain consensus that there should be a change in Bretton Woods. The bad news is that everybody wants a different kind of change.

The developing countries feel that the World Bank and the IMF have not been responsive to their aspirations and that they are not democratic in their working. Some nations, particularly one of our major donors, increasingly feel that we are not serving their bilateral interests very efficiently, that we are working too much for the developing world. So the changes proposed for Bretton Woods in the 1980s come from radically different perspectives.

Now, I am one of those who came to the World Bank in 1970 for a short period and then stayed on, and I have, somewhat uncomfortably, assumed at least two different roles: first, to run the policy formulation operation within the World Bank and, second, to be an outspoken critic of World Bank policies, both within and outside the institution. And I believe it is a tribute to Bretton Woods institutions and their capacity for change that I have survived so long and have seen a good deal of change in the last eleven years.

When I came in 1970, a lot of the stress was on the philosophy of GNP growth and on infrastructure projects. There was literally no direct emphasis on poverty alleviation.

Over the last eleven years, we have come so far from this emphasis that infrastructure projects receive hardly one-third of the total lending of the Bank. Two-thirds of its lending is devoted to sectors directly linked with poverty reduction programs in the developing countries.

When I came in 1970, land reform was regarded as a socialist philosophy and as subversive talk in the Bank corridors. In fact, one of the first things Robert McNamara faced as the new president of the Bank was a request from Chile to help with land reform programs. He was advised not to touch it. In 1974, when we sent our first paper on land reforms to the World Bank, McNamara related this story and said that, after a six-year period of learning about development problems, he was now making land reform a prerequisite in many countries for agriculture loans.

Again, before 1970, there was no lending to publicly owned industries. The ideology, at least, was that the Bank was to help private industries and not so-called publicly owned or nationalized industries. In the last eleven years, three-fourths of our lending in the industrial sector has been to public industries and to public development finance corporations.

There has thus been a certain pragmatic response by the World Bank to changing events. The world doesn't stand still, and any institution that stands still will be wiped out by the forces of change.

It will become irrelevant. The World Bank has managed to remain relevant to a changing, fast-moving situation by making a fairly rapid change in its own policy framework.

However, while the pace of change has been fairly rapid in the past, the World Bank and the IMF will have to undergo very major changes in their policies and structures in the 1980s if they are to remain relevant for the future. Let me elaborate.

The World Bank plays four different roles: it transfers resources to developing countries; it is committed to accelerating growth and reducing poverty; it contributes to the global development dialogue; and it has to respond to North-South issues, particularly in adapting its own institutional structure.

Insofar as resource transfers are concerned, there is a certain rumor around in the world that the World Bank is too large. Don't believe it. We are, in a way, a victim of our own propaganda. It's true that the World Bank expanded very fast in the last decade from a small bank. Mr. McNamara has made it into a very large bank, and there has been a fivefold expansion in Bank lending in nominal terms. But if you take away the effect of inflation and describe this growth in real terms, we have only about doubled our lending. Today lending amounts to about $12 billion from both IBRD and IDA. This is *less than one percent of the total development expenditure in the developing countries.*

Many of us in the World Bank often do not recognize it, but this is a sobering thought for any institution, and it indicates how softly we have to walk in the policy corridors of the developing countries where our lending is truly minimal.

We provide merely one percent of the total development expenditure in the developing countries. It's true that we do get a lot more mileage out of it by sharing various consortia and consultative groups, cofinancing with other sources of assistance, contributing to the international policy dialogue on a country through our economic reports, and lining up some assistance from private markets. Still, that all has to be kept in perspective.

Furthermore, the needs of the developing countries are much larger in the 1980s. There are at least four reasons why the World Bank needs to provide a much larger transfer of resources now.

First, China has become a member of the World Bank. This step alone has increased the population of our borrowing member countries by 45 percent.

Second, the developing countries are going through a major adjustment problem. They currently have deficits of $60 to $70 billion a year. This situation requires structural adjustments in the production, consumption, and investment patterns which need to be financed.

Third, a huge investment in energy, amounting to as much as $80 billion a year, is required by the developing countries.

And fourth, inflation is much higher now than was expected initially.

In light of these four factors, the World Bank's current transfers of resources to developing countries are very small in relation to what the needs are. And yet it is going to be extremely difficult to get additional resources to transfer through the World Bank.

We can do it through relatively costless techniques available to the Bank. We do not have to borrow from or put any burden on the budgets of the rich nations. We can recycle market funds by standing as an honest intermediary between private markets and developing countries, taking short-term loans and working them into long-term loans and assuming risks which commercial markets are unwilling to assume. We can do so by changing our present gearing ratio — that is, the ratio between our capital and our lending. It is 1:1 at the moment. In commercial banks, it is 25:1. In other words, they can lend $25 against $1 of capital.

Thus, there is a good deal of scope for enlarging our lending without really mortgaging our own credit-worthiness. There are other techniques available as well. We can have another capital replenishment without asking the developed countries to pay in any more. We can cofinance more with the private markets. We can use our guarantee powers a lot more.

The same cannot be said of IDA lending. IDA grants "soft loans" with only a commission charge of three-quarters of one percent for a fifty-year period to the poorest countries. The money to subsidize this form of lending comes from government contributions to the Bank. This source of Bank lending is always scarce, and there is pressure already for reduction in IDA "commitments" by donor countries.

So the dilemma of the 1980s is this: the possibility of a major imbalance in the role of the World Bank for middle-income developing countries and the poorest developing countries.

We might be able to increase our lending on hard commercial terms through the IBRD. We may be far more limited in our soft window (IDA) lending to the poorest countries. And that will create a major imbalance in the role that the World Bank has played in the past and the one that it has to play in the future, because the poorest countries are now in the greatest trouble.

There is also a certain worry, as I said, that probably the World Bank is too big. When people point that out, it is not in terms of need; it is more in terms of style of management. The issue is whether we should decentralize more, whether we should regionalize our operations — issues that we need to consider in any case in the 1980s.

My own favorite proposal is that if further expansion of the Bank's activities is to be done, it should come through new windows which should carry a different structure of control, so that there is a possibility both for expansion and reform. We can open new windows through which fresh air can come in.

If energy lending is to be expanded, a good way of doing it is through the establishment of an energy affiliate linked with the Bank, but which carries a different resource base, a different political structure, a different policy framework, and its own staff. Yet it would come under the general umbrella of the Bank in order to get started quickly vis-à-vis the financial markets.

We should be aware of the fundamental battle which is currently shaping up between multilateralism and bilateralism. The best days of multilateralism may be over. There was a certain period when multilateral aid through international institutions was regarded as superior to the state-to-state bilateral route.

Increasingly the pressure is coming from creditors who claim that international institutions are probably getting too internationalized in their approach and philosophy, and are not serving bilateral interests. I will illustrate one example of this conflict.

As you know, the official development assistance consists of concessional resources, the scarcest resources in the world. When they go through bilateral channels, only 34 percent of those resources currently go to the really needy, the poorest countries. Sixty-two percent are given on the basis of special relationships, and they go mostly to middle-income countries or other countries which do not require concessional assistance as much as access to international market opportunities for exports and capital borrowing.

When the same donors give their money to IDA, 92 percent of their resources go to the poorest countries. So basically the edge that multilateralism can claim is that of channeling these resources according to poverty needs and according to productive use of funds. It cannot claim that these resources will be distributed in accordance with the foreign policy interests of donor countries. This is a fundamental conflict that I will come back to later.

Second, the Bank's role in development and poverty alleviation. As I was saying, the World Bank has changed its role considerably in the 1970s. Its focus is on increased productivity of the poor and also on the provision of public services in the developing countries, alongside the traditional sources of economic growth.

In this new role the Bank chose to stress productivity aspects rather than poverty aspects to indicate that this assistance is not charity or "money down a rat hole." Thus, the poorest are able to increase their own productivity and become self-reliant. That's why the Bank finances programs through which small farmers can increase their own productivity by credit-extension services and other delivery systems that we put together along with the government.

These sectors increased in our lending from about 8 percent in 1970 to 30 percent in 1980. The rate of growth of lending in these sectors has been four times the average rate of growth of our total lending.

But in the 1980s, there is going to be a very major debate within the Bank, and outside it, on the Bank's focus on the poverty issue, and there will be many attempts at revisionism. This revisionism will assert itself partly because some of our contributors are changing their own emphasis on the poverty issue within their own domestic spheres, and it is logical to expect that they will want to see the same kind of emphasis reflected in the World Bank.

There will also be a major fight over this issue within the Bank. McNamara, through the force of his personality and his very strong leadership, was able to get some of the ideas accepted, but it is very difficult to undertake poverty projects successfully.

It is not very glamorous to walk in the villages to get to the small farmer, to devise new delivery systems, to undercut the existing power structure within these systems. The glamorous projects lie somewhere else. They are generally in the urban areas, the industrial turnkey projects. A very major battle in the coming decade is going to concern the focus of the lending policies of the Bank.

Furthermore, let me also say in all candor that the poverty problem is one we have just begun to address. We are just beginning to learn how to deal with this issue in the developing countries, and there are at least four areas in which we are very weak.

We are very weak in working with those people in developing countries who have no assets. A small farmer has some assets, and you can somehow work alongside the government to increase the productivity of those assets. But what about the urban poor? How do you reach a shifting, restless mass? Or the landless labor? Our programs and policies still have not found a satisfactory answer.

Similarly, we have done a poorer job of reaching the bottom 10 percent of the population in developing countries than we have in reaching the top part of the bottom 40 percent. To reach the poorest and to sustain them for a certain period until they can be brought into the mainstream of economic development often requires some subsidies, and we have not been very tolerant of subsidies for understandable reasons, because they can become perpetual and undiscriminating and, in fact, often have. Thus, the Bank has had limited impact on the bottom 10 to 20 percent of the population in developing countries.

We have generally financed capital expenditures, such as construction of schools, buildings, and hospitals, but not the salaries of teachers and doctors. This is an all too familiar phenomenon in the developing countries: to find schools without teachers and thus wasted infrastructure investment. However, we have recently changed our policy on this, and the Bank will now also be financing recurrent expenditures in education for a certain period of time, with a commitment by the government that eventually it will take over this expense.

At the same time, another issue has come up: How do we consolidate our emphasis on poverty programs on the one hand while moving to meet the new needs which have arisen in energy, industrial production, and restructuring of entire economies on the other? This is a very cruel dilemma for the developing countries; it is for us, too.

Our response to this problem is that we should have additional resources to do both, but we are not getting these additional resources. Choices are being made in response to the immediate short-term crises, often at the expense of some of the longer-term social programs, a choice not to be recommended.

The Bank also plays a role in the global development dialogue. The perception outside the institution is that the World Bank has not been very responsive to South-South issues and that its world development reports carry a certain flavor of the Chicago school. These reports advocate the price system and its application to the problems of developing countries. The most recent report, however, which was devoted to basic needs and human and social development, helped correct some of that impression.

My own view on this is that the World Bank can clearly play a major role in professional clarification of global development issues.

If there are issues like food security or energy security or any other issue on the global agenda, there is no reason why we should not be doing the professional analysis of various policy options that the international community faces. We can't be prescriptive because of the different schools of thought among the staff and the board members. But we do owe it to ourselves and, I believe, to the global community to clarify policy choices on a number of global issues, without necessarily offering a final set of policy recommendations.

Let me now turn to the final issue of the Bank's structure, which excites a number of people inside and outside the Bank.

When the World Bank was created in 1946, the prevailing balance of power was naturally reflected in the Bank's structure. The developing countries were represented at the Bretton Woods conference mainly by India (for Asia), which was a colony at that time; by Ethiopia (for Africa); and by Brazil (for Latin America).

Surely, if there were a Bretton Woods conference today, there would be a different cast of characters sitting around the table. Even the balance of power between the United States and Europe was very different at that time. Europe was just emerging from the rubble of World War II. It is difficult to imagine today that Germany and Japan were given a combined voting power lower than that of the United Kingdom.

There are three issues that have arisen, generally raised by the Group of 77, UN global negotiations, and the Brandt Commission: the management structure of the Bank, its voting structure, and its membership structure.

The management structure should have been the easiest one to handle, because there is no political quota in the World Bank, unlike the United Nations. We at the Bretton Woods institutions take pride in the fact that we recruit people on the basis of their professional caliber, irrespective of nationality.

It is very difficult to understand, therefore, why still only 17 percent of the World Bank's senior management is from developing countries, which is even less than half of the political quota the developing countries have on the Board. And this question of management bothers the developing countries a lot more than even the issue of voting structure, because they are more exposed to our staff for lending operations than to our Board. It is true that we started with a very small base of developing countries that has been expanding over time, but I am not convinced that the management structure cannot or should not change drastically in the 1980s.

Second, on the voting structure I am more pessimistic, because there is a fatal flaw in the conception of the Bretton Woods institutions. They are called international institutions, but they are not.

They are really an uneasy coalition of bilateral interests. They are a shareholders' company, and if the borrowers wish to have majority control, the lenders will take their money and go away. I have not seen a single bank or a single lending institution, national or international, where the borrowers are in majority control. So the only way the voting structure can change in favor of borrowing countries is if they — the developing countries — become lenders to the institutions, because votes are in relation to the capital that they subscribe (except for a few membership votes).

I believe that gradual steps should be taken to "internationalize" the Bank's resources. Unless resources are internationalized, the boards of institutions can never truly be internationalized. It is a fallacy to think that institutions impose the constraint themselves. They do not. The shareholders do. And unless we have future institutions based more on automatic mobilization of resources, on international taxation, on resources from the global commons, on the SDR link, or whatever — unless we internationalize the resources — the control will not get internationalized.

And the third question is membership. The World Bank still excludes a very large part of the international community, particularly the communist countries.

Of course the communist countries opted to stay out of Bretton Woods in the 1940s; they were not pushed out. They regarded these institutions at that stage as creations of the Western powers.

The question is whether, in the 1980s or 1990s, one-third of the international community is going to stay out of institutions that are so vital for resource transfers and international monetary policy, especially when socialist countries are coming out more and more into the international community by way of trade and by way of their monetary dealings. The option of following the bilateral route or barter route will no longer be there in the next two decades. This is another dimension that has to be incorporated in thinking about the future of Bretton Woods.

Now, how do we do it all? One suggestion, based on the belief that fundamental changes are needed, is to hold another Bretton Woods conference, maybe under UN auspices. But people fear that if there were such a conference, it might continue in a state of paralysis and yield no results for a long time. Therefore some people suggest marginal changes that might, over time, accumulate.

I'm not quite sure how the changes will come, whether they will be the result of a crisis of multilateralism, a crisis in IDA resources, a crisis of the international monetary system, or a crisis in IMF. But I do believe that the changes in Bretton Woods structures are of a very fundamental nature in the next decade. We can move toward them gradually by opening new windows and in that fashion try to incorporate the new balance of power. Over time, maybe that will exercise enough pressure on the total institutions. Or we can try to do it through another Bretton Woods conference. But these are the issues we are all going to live with in the 1980s.

Finance and Development

Paul Streeten: I want to begin by referring to an interesting little booklet by Harlan Cleveland, former Distinguished Visiting Tom Slick Professor of World Peace. It was published in 1977, and its interesting title is "A Third Try at World Order."

I wonder what those of you who haven't read it think the title would cover? What would we think of as the three tries at world order?

Harlan Cleveland very interestingly describes the League of Nations as the first try; the United Nations as the second try (we all know the inadequacies and failures of the UN), and the challenge of the third try is: Where do we go from here in our present interdependent world?

That's an interesting analysis, but if I were to write a book under that title I would deal with three different tries. I would analyze world order in terms of three stages, going somewhat farther back in history than Harlan Cleveland did.

The first try was essentially what is called the "Pax Britannica." The second try was the "Pax Americana." And the third "nontry" is the present kind of world disorder, schizophrenia, or pluralism, or whatever you would like to call it.

Under the two great peaces — the peace imposed by the British Empire and the peace imposed by America — four types of activity were concentrated in one super power. It was one power (Britain in the nineteenth century and until 1914 and America until 1970) which generated surpluses in its balance of payments which then had to be in some way financed; which produced and supplied the capital goods purchased abroad with these loans; and which evolved the required financial institutions. The same dominant country that produced the surpluses and the capital goods controlled the financial institutions that channeled the funds, so balance-of-payments surpluses, financial institutions, capital goods, and other types of goods which promote development were concentrated in the same place, essentially London until 1914, and New York and the United States up to, say 1971.

Moreover, the seat of military power and military strength, on the whole, was concentrated in one place, which in emergencies and crises backed up the orderly system. Military power went with economic power.

Today, and in the last ten years or so, however, the balance-of-payments surpluses have tended to be generated by more than one country, above all by some OPEC countries. Occasionally one or two OECD countries also generate surpluses. Japan and Germany have done so, and they may do so again. So, it's not purely OPEC countries which are surplus countries.

The countries that produce the capital goods are, again, a number of different European countries and Japan. The financial institutions are still largely in London and New York. The economically strong — Germany and Japan — are not militarily strong. And the military superpowers are economically ailing. As a result, we have a form of pluralism — or perhaps monetary schizophrenia would be better: a division of activities and functions for which we have not yet evolved the appropriate institutional coordination.

One problem of evolving an appropriate new international order is to create the institutions that deal with a world in which three or four important functions are divided among different countries of different types. In the light of its own objectives Bretton Woods has functioned very well for twenty-five years in a world in which there was a Pax Americana, with its drawbacks and advantages.

But in this new pluralistic world with a fragmentation of these functions, we have not yet evolved the right institutions.

In an interesting recent book, the French writer Servan Schreiber says that the oil sheiks have the capital, and they don't know what to do with it. The advanced industrial countries have the machines they do not know what to do with. The developing countries have a large reserve of labor, and though they do know what to do with it, they haven't got the capital or the machines to provide a minimum standard of living for their people. The world has not been able yet to bring these three sources of productive development together. It calls for an imaginative response.

Now, let me turn very briefly to the role of the International Monetary Fund in this. I want to make three main criticisms, and then a number of proposals for dealing with them in the limited area of monetary reform and short-term lending. These remarks will complement what Mahbub ul Haq has said about long-term lending to the poor, and what Gerry Helleiner will say about trade, about which I shall say nothing.

The IMF has been quite successful up to a point. What Mahbub has said about the Bank can also be said about the Fund. It has adapted, it has changed, it has adjusted to the new challenges in a quiet, innovative way, but not rapidly enough in the light of changing circumstances and changing objectives.

For example, the Fund now deals with the balance-of-payments deficits and surpluses generated by the OPEC surplus countries as if they were essentially due to domestic mismanagement, to excess supply of money, or to the wrong kind of exchange rates. Now, it seems to me that the way to deal with the effects of OPEC surpluses must be fundamentally different from the traditional ways of reducing the supply of money or adjusting exchange rates and so on. As a result of this wrong analysis, the IMF tends to impose the wrong conditions on the countries borrowing from it.

Linked to this but separate is the second criticism that the Fund, perhaps because of the way it is set up, does not take a global view of global issues of monetary reform, but a country-by-country view. There are certain situations in which a country-by-country approach produces worse results than a global view. If a group of countries is generating surpluses, somebody has to accept the deficits. If everybody tries to get rid of the deficits and kicks them around to the others, we get into the situation we are in now, where we have either "beggar thy neighbor" deflation, a policy that tries to reduce demand in incomes and employment in order to get rid of the deficit, or we get competitive protection if we try to prevent imports. But somebody has to hold the "hot potato," and this does not seem to fit into the present philosophy of the Fund.

The third criticism is that the resources of the Fund are far too small. In one recent year, far from disbursing money, the Fund actually absorbed money, and in 1979 it only disbursed the financial equivalent of 0.16 percent of world exports. So one would want to increase the resources of the Fund.

Now, very briefly, to some specific and pragmatic proposals for how to improve this situation and what to do about the Fund. I would suggest, although I cannot discuss them in detail, six proposals which I think would move us in the right direction and do something to adjust the institutional lags created by our newly pluralistic world.

First, I think most of us would agree that an important and serious move toward the establishment of a new international currency is needed. In the Pax Britannica, the international currency was gold and sterling. In the Pax Americana, the international currency was the dollar. Neither of these is suitable anymore, because they fluctuate in value. We want to move toward a genuine international currency that maintains its real value in terms of other currencies. That, of course, is the so-called special drawing rights (SDR), which is still very small and not issued in adequate amounts. Or perhaps, as some people have suggested, we should move toward what is called a commodity standard, which would maintain its real value not only in terms of other currencies, like the SDR, but also in real terms, backed by a bundle of commodities.

The second proposal concerns what is called in technical jargon a "recycling mechanism," an improved mechanism by which we transfer the surpluses of the OPEC surplus countries and some OECD countries to those countries capable of accepting the deficits. That in turn would mean increasing the quantity that is recycled, so that we avoid competitive deflation or competitive protectionism. We must reduce the uncertainty for the surplus countries, thus encouraging them to invest their financial surpluses. This is the kind of proposal that Professor Helleiner made yesterday when he began to discuss the recycling program.

It also means a change in the country distribution of the lending of these surpluses. It is always said that bankers only lend to those who need it least. The money is recycled and lent, by and large, to the countries that are really quite well off — to the middle-income countries, the Brazils of this world — and not to those poor countries who need it most.

To avoid debt crises, we should lengthen the maturities of some of the debts. We need some innovation in creating a new type of long-term international bond or an agreement between central banks to on-lend or an international investment trust. These are aspects of improving the recycling mechanism.

A third proposition — which has come up in the Brandt Commission report and which some of us would support — concerns some new lending facilities for medium-term program lending. By "program lending" we mean loans that are not for specific projects, but the kind of thing that the Fund does: balance-of-payments support loans but for much longer periods than three or six months — perhaps for twenty, thirty, forty, or fifty years.

A fourth important change I would advocate and one which arises from my criticism of the Fund concerns the way that the Fund operates and the kinds of conditions that the Fund imposes. We would like the Fund to be more sensitive both politically and technically to different ways of doing things and to a broader range of alternatives.

By and large, the basic objective of the Fund is monetary stability — to prevent or correct inflation and balance-of-payments deficits. But, of course, there are other objectives than these, and one would like these other objectives to be taken into account. And it should permit alternative ways of handling inflationary balance-of-payments deficits, other than exchange rate changes, public expenditure cuts, and reductions in supply of money. Inflation maybe cured by stimulating supply as well as by restricting demand. There should be a greater awareness that objectives and instruments can be more varied than the Fund permits (e.g., by allowing direct controls).

The other point that has of course been made for many, many years is that the present system is very asymmetrical. Very heavy pressure is exercised on countries that have balance-of-payments deficits. They import more than they export, and they have to take some measures to tighten their belts, but hardly any pressures are exercised on the surplus countries. Keynes and many of us since then have emphasized that a balanced international system should use some pressure on the surplus countries to either lend their surpluses on long term or to reduce their surpluses by buying more from other countries so that this kind of asymmetry, which presents a deflationary bias, is reduced.

As I said before, the Fund should change its country-by-country approach to a more global approach, to see the world as a whole and to realize that policies which are appropriate for a single country are not necessarily appropriate for all countries together.

A fifth suggestion concerns the present system of floating exchange rates. The system of floating rates, on the whole, has been quite disturbing and upsetting to many developing countries, and there is a need to soften its impact in these countries.

Let me mention measures which have been much discussed in the literature, such as controls on short-term capital movements, more limited variations of the exchange rate (what, in the technical jargon, is called "crawling pegs"), and various others. Floating rates have not fulfilled the hopes. I have never thought that they would, but many of my economics colleagues said, "Once we have floating rates, we will liberate national policy and be able to achieve all of our objectives without worrying too much about the rest of the world." They have not done that. Floating rates have brought their own problems with them, and we should think of ways of dealing with them and preventing their disturbing impact on the developing countries.

And now a sixth point. What is needed to strengthen the developing efforts of developing countries is finance for South-South trade and South-South cooperation, that is to say, for cooperation and trade among developing countries.

The international institutional network is still largely North-South — from the North to the South — in terms of lending and in terms of flows of trade and capital goods. Finance is needed to establish trade among the developing countries.

And finally, the whole situation with respect to gold has changed enormously. The IMF now has 100 million ounces of gold. While the Fund has sold some of this gold in the market and used the receipts for development purposes, much more can be done by way of evolving gold accounts and by way of correcting one of the deplorable results of this monetary schizophrenic system we have been living in for the last nine or ten years. Far from creating an international currency base in SDRs, we have created an enormously large increase in the reserves of those countries that happen to hold gold, or those countries that produce gold, because of the fantastic increase in the price of gold in the last ten years. There seems to be, again, a very uneven and arbitrary increase in international reserves, and we should do something to correct that. Perhaps we can compensate the developing countries for their not fully noticed relative losses. This compensation could be based on the share in international reserves following from the increase in the price of gold compared with the situation in which the distribution of international liquidity would have been more rational according to some set of universal principles.

We will now turn to Professor Helleiner who will speak on trade and development.

Trade and Development

Gerald Helleiner: The trade issue is the issue par excellence of mutual interests. It is the one which economists, for years and years, have been preaching about — that all stand to gain from a mutual exchange of goods and services. Yet the issue is one in which particular interests, expressing themselves through the usual political channels, have consistently stalled and made more difficult the progress which is in the global interest, at least according to economic analysts.

The "accommodations" that we are to discuss this morning are of two kinds. One is the short-run need simply to observe the rules, to restore some credibility to those institutions, frameworks, norms, and standards which we thought we were establishing in the period immediately after World War II.

The global trading system has been in some respects rotting away, and the principal costs are being borne by the weakest, as always. Rules are, in this instance, of greatest concern to the weakest, who would otherwise be bullied by the powerful into outcomes which are the product of no rules whatsoever.

The second kind of "accommodation" concerns long-run issues. These are rather broader and have to do with the capacity of the existing institutional machinery, even if it ran properly — and there is sand in the machinery right now — to handle the emerging range of new world problems in the trading sphere. I hope to talk about both the immediate and the longer-run issues in the trade sphere.

Let me first underline the importance of trade to the developing countries. It isn't always adequately realized that the sheer value of nonoil exports from the developing countries is ten times that of the total official development assistance flows to these countries. It's over three times that of total resource flows of all kinds, including direct investment, commercial bank finance, and so forth. In 1980 total nonoil exports amounted to $250 billion.

The dimensions of the trading issues, then, are enormous. It's very important to get them right.

At the same time, the importance of developing country markets to Northern exporters has frequently been underestimated. By now, these markets ought to be a major political fact of concern to Northern exporters and Northern banks who want to be repaid, having lent substantial sums to countries which are now experiencing difficulties selling the exports in Northern markets, this being the only way they can earn the foreign exchange to repay the debt.

As a political matter, it's a continuing mystery to me why exporters and banks do not fight against the particularistic groups which are impeding market access for the developing countries. Perhaps we will hear more in this morning's discussion about these sorts of political factors.

For the United States, the nonoil developing countries now account for more exports than all of the European Economic Community, plus all of the centrally planned economies, counting China and the USSR as well as Eastern Europe. And United States exports to nonoil developing countries are over three times those to Japan. These are not insubstantial interests which are at stake for American exporters.

The composition of developing country exports changed extremely rapidly in the 1970s. In 1979, for the first time, manufactured goods surpassed primary products in their importance in the export earnings of the nonoil developing countries.

I could say many things about primary products trade, about agricultural protectionism, about the systems of marketing, distribution, and processing in the minerals and primary products sphere. But I think it more important to concentrate on the manufactured goods trade, which is the main subject of the increasing concern about protectionism.

These new manufactured exports are coming not only from the newly industrializing countries, but increasingly from a wide variety of other developing countries, most recently a collection of countries in Latin America. There is an inevitable, unstoppable process of economic structural change underway at the global level. The Third World is industrializing. There is a change in the global location of particular industries. Investment decisions are being made at the global level in a way profoundly different from how they were made thirty years ago.

Despite this process, manufactured imports from developing countries still account for a remarkably small percentage — about 2 percent — of the industrialized countries' total consumption of manufactured products. For all the talk, you would think it was rather larger.

Even in textiles and clothing, the developing countries only account for 5 percent of total OECD textile consumption, and about 10 percent of total OECD clothing consumption. In fact, the OECD countries still account for over half of the world trade in both textiles and clothing.

In the case of clothing, about 70 percent of total world trade is still industrialized country trade. Yet, in the last part of the 1970s, protectionist pressures began to emerge within what is supposed to be an open, liberal, nondiscriminatory framework for international trade, the framework established by the GATT.

The main principle of the GATT was and is nondiscrimination. All countries are to be treated equally. What has emerged under the so-called "new protectionism" is a set of discriminatory practices, some of which are completely contrary to the GATT itself, some of which are not covered in the GATT, and some of which are authorized by special exemptions and protocols under the GATT. But all of these are having the effect of hurting those developing countries least able to retaliate, least able to undertake any sort of reciprocal self-protecting action.

There has been a burst of nontariff barriers of various kinds, many of which are very difficult even to write down on paper. They involve a "nightmare of confusion," as one American commentator called them. They involve large numbers of lawyers and a need to gain access to the appropriate authorities. They don't cause as much trouble for American-based firms importing into the United States as they do for foreign companies trying to do so. There are thus implicit biases in the way the new protectionism works. All of this has created substantial new uncertainties for investors.

It's very difficult now for developing-country planners and investors to know what to expect. There is an uncertainty about what can now be expected concerning access to Northern markets if investments are undertaken in developing countries. And what the developing country group consistently asks for is a reduction in at least the "manmade" uncertainties. The are faced with droughts; they are faced with global price fluctuations in their commodities; they are faced with recessions. They are faced with all sorts of, so to speak, "natural" uncertainties. But to be faced, in addition, with discriminatory trade policies undertaken at their expense is the final straw.

In the short run, there are three main issues that need resolution. The first relates to the major failure of the Tokyo Round of bargaining concluded recently under the GATT. The Tokyo Round did succeed in reducing tariffs on a number of items and did succeed in writing codes of behavior with respect to certain nontariff barriers among the participating countries. But one major failing may be more important than all the progress.

Under the GATT there is one major safeguard clause, Article 19, which authorizes countries, in certain circumstances, to impose import quotas against other members of the GATT. That clause strictly requires that, when a country imposes import quotas, it does so only because there is serious injury or threat of injury in the industry concerned. When a country imposes these quotas, it is to do so on a nondiscriminatory basis, maintaining the nondiscrimination principle of the GATT. The country is also supposed to consult interested parties in advance.

These provisions of the safeguard clause were both too strict and too weak. They were too strict in that the nondiscrimination required made it very difficult for importers to pinpoint their action and direct it against those countries that were causing them the most trouble; they were too weak in that there were no provisions for international monitoring or surveillance, no requirements for domestic adjustment, and no time limits on the imposition of these restrictions. The result was that increasing numbers of countries ran around the end of Article 19 and developed nontariff barrier systems which did not operate under the GATT norms, but operated under no norms whatsoever. In particular, if one strong nation, say the United States, can, through diplomatic pressure and bullying, infuse a weaker country to "voluntarily" restrict its exports, then this action is not covered under the GATT. Furthermore, if an importing nation can, through joint arrangements with exporting countries, organize "orderly marketing arrangements," in which the trade at issue is managed through cooperation between governments, this is not covered by the GATT either.

This method of circumventing existing GATT provisions has, in fact, provided enormous power to the major industrial importing countries to do essentially whatever they want, and that is precisely what they have done in an increasing number of areas.

The Tokyo Round of talks totally failed to rewrite the safeguard clause, and the highest priority in the short run must be to get that safeguard clause in a form which reins in unilateral bullying tactics on the part of the most powerful trading nations.

Second, the textiles sector was long ago made an exception to the GATT norms. In 1961 a so-called "short-term arrangement" was negotiated as part of President Kennedy's effort to get his Trade Expansion Act through the United States Congress. He induced the members of the GATT to make textiles an exception and to authorize, as a temporary measure, discrimination against low-cost textile exporters, now located almost exclusively in developing countries.

In the following year the "short-term" arrangement became a long-term arrangement. Originally it covered only cotton textiles, but by 1974 this arrangement had been extended to all other fibers and had become a "multifiber" agreement. At that time there were certain limits on it. The authorized quotas were to grow by 6 percent each year. The quotas were not to be any smaller than at least the imports of the previous twelve months. They were not to be imposed against very small new entrants to the market. The agreement contained far more checks and balances than did the last extension of the multifiber arrangements, which was made in 1978 and expires in 1981.

In the most recent protocol of extension, a crucial phrase was written in, authorizing the importing countries to do essentially whatever they choose under certain circumstances. The phrase was "reasonable departures." Reasonable departures from the original multifiber arrangement were authorized, and those reasonable departures have now become the norm.

The sizes of the quotas have been rolled back. The growth rates in the quotas have not followed the 6 percent rule. The requirement that market disruption be demonstrated has been abandoned. The international surveillance system is a joke. When the agreement comes up for renewal this year, my expectation is that the developing country group will refuse to sign it. When they refuse, it will force the pressure back onto the GATT to rewrite the safeguard clause. Textiles will no longer be the exceptional case, to be handled in this unique fashion. The developing countries know that footware, automobiles, steel, shipbuilding, electronics, and a lot of other industries are coming along, and some way has to be devised for handling each of them. There are going to be more and more of these industries and these developing countries have had enough, I think properly so, of this "special case" treatment.

Lastly, the codes tortuously negotiated under the Tokyo Round of talks introduced a new phenomenon in the GATT, discrimination against those who do not sign particular codes. The codes which were signed with respect to customs evaluation, the right to impose countervailing import duties when a country subsidizes its exports, and so on apply only to those who sign the agreement. The developing countries, almost without exception, have not signed the agreement because they didn't get a decent safeguard clause agreement. Developing countries are therefore unlikely to derive many benefits from the widely heralded new codes.

There are many other aspects of the Tokyo Round which were unsatisfactory. A promise was made and widely announced at its outset that there would be special and preferential treatment for developing countries. It did not happen. Since the key item of interest for developing countries failed, it's not surprising that they don't sign.

But now, confronted with a code system which applies to them only if they do sign, they are, in effect, being discriminated against once again. Countries like Mexico don't belong to the GATT; there are forty or fifty other developing countries that don't belong. This is one way in which the GATT is unlike the IMF and the Bank. These nonmember countries will not benefit from anything negotiated on the nontariff barrier front anyway. Even those who do belong and do sign the codes now have reason for concern, because there is an escape clause in the crucial countervailing duty code which permits the importing country to avoid the demonstration of injury that would otherwise be required to authorize countervailing duties in instances where the exporting country subsidizes exports; and the United States is among those countries which are now busily getting the formal right to avoid the code on countervailing duties, so that they can handle those circumstances where developing countries like India are prepared to sign the code.

Therefore, in the short run, there are all kinds of problems in these three major areas: getting the codes right, getting the safeguard clause in place, and getting the multifiber arrangement handled one way or another. These are areas of immediate concern, and immediate solutions are needed to restore some order.

What really deserves much more attention is a long-range problem: the entire GATT system is creaking. It's not a matter of just stopping the rot, which is very much an immediate concern, but it's also a matter of having the institution catch up with the times.

Paul has already mentioned some of the aspects of global trade which are of increasing importance but which the GATT does not cover: intrafirm, intracorporate trade; state trading; agricultural trade; and about all, restrictive business practices, which are peculiarly handled in another UN institution entirely, the UNCTAD.

The main trouble with the GATT is that it carries so little credibility among the developing country group. It only has 86 members. The United Nations has over 150 members. As I mentioned, Mexico doesn't belong to the GATT and many other developing countries don't belong to the GATT. The institution is simply not fully trusted. It is not seen as something that offers the developing countries significant benefits, particularly when nobody observes the rules.

At the same time, the UN Conference on Trade and Development (UNCTAD) takes a keen interest in trade issues and does have credibility among the developing countries, but suffers a credibility problem on the part of the industrial countries. It handles some issues, like restrictive business practices and primary commodity trade, which are clearly trading questions and which have substantial overlap with those discussed in the GATT. It makes very little institutional sense to have these discussions proceeding in two different places, and it creates mutual suspicion on the part of both institutions and their member governments as to what each of these two trade institutions is doing at any one time.

My final point about long-run concerns is that we need a "beacon" in the trade sphere in the very worst way. We need to know where we're going, what we're trying to achieve a long way down the road, and where we want to be, because there is a very long gestation period for even minor reforms. It took seven years to achieve the Tokyo Round results. It will take longer to fully restructure trade institutions.

Yet the global economic system is under very great pressure. We are in a period of slow growth. We are in a period in which the Third World is industrializing and entering the world market rapidly with many new industries which will be competitive with Northern industries. We also have rapid technical change — the computer and chip revolutions — causing further adjustment problems. There will be great pressures on employment and political pressures for protection on trade policy makers in developed countries.

In the 1980s, if we do not work persistently toward an agreed-upon framework, an institution that in some way integrates the various trading concerns and puts them in one place, and a set of rules that really works and, in particular, protects the weakest, we will slide backward. We have usually slid back even in reasonably good times because of the pressure of political interest groups expressed through the usual parliamentary and congressional systems.

In bad times, the degree of disorder in the world trade scene, without a beacon of some kind, will be very great. I very much hope that the Western Economic Summit will soon erect such a beacon to which both Southern and Northern leaders can look at the North-South Summit and over the course of the 1980s to restore some credibility to the international trading institutional machinery.

Chairman Streeten: Thank you very much, Gerry. The discussion is now opened to members of this panel.

Panel Discussion

Chairman Streeten: Professor Weintraub.

Sidney Weintraub: First let me respond to some of Gerry Helleiner's points. He spoke as though the world trading system is rampant with protectionism and that trade is being impeded at every single instant. I would like to challenge that perspective.

I start out, as he does, as a free trader. I think that the best of all worlds would be a world without restrictions of any kind in developed countries, and I think that's the world he starts with. At the same time, we live in a world of compromises, and he left the impression that all of the compromises have been on the protectionists' side.

I think it's a pretty false picture. The danger is present, but it's still only a potential danger. It's not yet the picture that he painted. Trade growth, particularly the area of trade manufacturers, has been exceedingly great throughout the post–World War II period. It slowed down somewhat in the mid-1970s partly because of the stagnation in the industrial countries where the markets were. But the developing countries, particularly those that have manufactured goods to export, have grown faster in GNP percentage terms than the industrial countries.

Gerry Helleiner pictured the United States market as reasonably closed. We take more manufactured goods from developing countries than does any other country in the world. He painted a picture of restrictions on the poorer countries like South Korea, Mexico, Brazil, Hong Kong, and Taiwan. Yet those are the countries that sell about 80 to 90 percent of the South's manufactured goods, and they haven't been impeded more than marginally. In the 1960s the quantum growth of their exports of manufactures was about 30 percent a year in some cases, 20 percent a year for others. And in the 1970s, for the most part, this growth continued.

After the great oil crisis of 1974, when most of us in the economics profession were deeply, deeply concerned that there would be a reversion to beggar-thy-neighbor trade policies because of the slowdown in growth and the tremendous balance-of-payments deficits that ensued, the OECD countries pledged not to impose trade restrictions on balance-of-payments grounds, and for the most part they have kept the pledge. I don't want to defend the cotton textile or the multifiber arrangement, but I don't want to leave the impression that it represents the norm. Most of the voluntary restraints that have been cited, other than the multifiber arrangement, have affected the large developing countries that I have just mentioned. They are the ones that have been affected by it primarily. And, again, I find it hard to accept the notion that South Korea, Taiwan, and Hong Kong have been unable to trade in manufactured goods.

Let me just take a second to trace what's happened in trade policy, because I think it's a record that we ought to be proud of, given the nature of the real world, rather than a record that we ought to be ashamed of, even though I'm a free trader.

The United States, in the Pax Americana period to which Paul

Streeten referred, set the tone for a series of negotiations reducing restrictions in international trade. The fact that tariffs are now negligible on most products in most industrial countries is the result of that initiative. The reason that we're focusing now on nontariff barriers is that we have succeeded in eliminating practically all of the seriously restrictive tariff barriers. The United States deserves more credit for this, I think, than any other country.

Helleiner wants us to negotiate immediately for a safeguard code in the GATT. I hope we don't. If we start now under this administration and with slow growth, the horror story would be so terrible I would rather not contemplate it. The safeguard negotiations in the Tokyo Round negotiations at Geneva were not successful because the United States wanted the safeguard code to maintain the principle of nondiscrimination, and most of the Europeans did not want to maintain that particular principle. They wanted to restrict imports in a discriminatory way against individual countries, and the United States held out against it. I'm afraid if the United States were to enter into a negotiation now, under the current circumstances with this administration, we would end up on the European side.

One other point I would just like to touch on: countries that have not joined the GATT.

Mexico did not join the GATT not only because of lack of credibility of the GATT. What Mexico was afraid of was that there would be pressure over the next two or three *decades* for Mexico to liberalize its own import licensing system more than it's quite prepared to commit itself to right now.

Let me put it in the bluntest terms: Mexico is a free rider in the trading system. Mexico gets most-favored-nation treatment without contributing and *quid pro quo.*

I guess I am bothered by a picture that paints everything black when the system has really worked better for the newly industrializing countries, if you just look at the growth in their trade, than any system actually has in any other period of trade history.

Let me mention one other thing to highlight the contrast between those who think in terms of gradual concrete accomplishments and those who think in terms of what the best of all worlds would be. I really have in mind Paul Streeten's comments on the monetary system.

He listed a series of proposals which would be useful in the monetary system, and I essentially agree with him. I don't disagree that you need more pressure on surplus countries, including the OPEC countries, and the privilege many of you suggested that we should give them is a guarantee of the value of their continued surpluses. Also, you want to make the foreign exchange system work better, and I agree that it should work better. You want better use of IMF goals, and that's very nice. You want a new international currency, and I guess a lot of people want this.

I don't want to be facetious, but there was a column in the *New York Times* not too long ago in which a man asked a series of questions, and then he answered the questions:

"What kinds of changes do we want?"

"We want fundamental changes."

"What kind of proposals do we want?"

"We want imaginative and innovative proposals."

"What kind of order do we want?"

"We want a world order."

"What kind of world do we need?"

"We need a just world."

"What kind of adjustment do we need?"

"We need a structural adjustment."

In a sense, I think that's what Paul Streeten is saying. For about ten or fifteen years in the developed and, to a certain extent, in the developing countries, economists have argued that the special drawing rights should become a world currency in the place of national currencies.

This idea really came to a test during the negotiations of the Committee of Twenty. Most of the resistance came from the developing countries, who did not really want the SDRs to be a world currency. They didn't want to lose the right to be able to hold their dollars and invest their dollars wherever they wanted to.

What Paul has outlined is very nice, but we (the nations of the world) are not quite prepared to do it. So while in theory I agree with him, there must be some reason why the practical people who have to deal with these reforms, including those from the developing countries, are not quite prepared to accept them.

I don't know how to make the exchange rate system work better. As a matter of fact, I'm impressed that, in this period of tremendous turmoil, the exchange rate system has actually worked as well as it has, and that trade has not really been impeded by the exchange rate system. It may have been impeded a good deal by the tremendous change that took place in the whole balance-of-payments structure around the world.

In theory I was one of those people who argued in advance that complete floating was by far the best thing that could happen, that free floating would free monetary policy internally. It would end protection, because a country would not have to worry about its balance of payments. We all know a little more now. We know that's not the way the world is, that free floating is not acceptable to most countries, that our nicely spun theories are now about ten years out of date because we have now had about ten years of experience with them. And I will revert to what I criticized you for: we really need some new theories now.

Chairman Streeten: Thank you. Gerry?

Gerald Helleiner: Please let me respond directly to Sid Weintraub. I am not talking about the 1960s or the 1970s. I thought I made that clear, and I certainly don't remember making any attacks upon previous United States policy.

I'm talking about the climate of the 1980s. And as the chief economist of the GATT himself put it at a meeting I was at about three weeks ago in Geneva, what is quite apparent now is that the members of the OECD are not prepared to treat the developing countries, and quite possibly Japan, as free and equal trading partners. That's the problem. And until a system is developed during this period of great stress — again, I'm talking about the 1980s — that encourages adjustment to processes now under way, the system in the 1980s is at great risk. It doesn't help to fight last year's war.

Of course, as I said, one of the reasons things are where they are is the great successes of industrialization that the developing countries achieved in part through the penetration of Northern markets. The penetration has been unbalanced and is still very small, so presumably there is lots of room for more. But the very uncertain climate at present is such that the developing countries' own investment activities are being curtailed. They cannot now decide what to do. It's a period of great uncertainty.

I was interested in the remarks about the present United States administration, because it had been my impression — but I stand to be corrected — that it is ideologically correct for this kind of administration to pursue free trade and nondiscrimination.

I think it's ideologically correct at the same time to favor international antitrust and agreements for the control of restrictive business practices, which I believe must go along with free trade if there is to be a serious attempt to improve global efficiency. I'm not a total free trader, unless with free trade goes the control of market power and its abuse. At least as far as free trade is concerned, it seems to me that the Reagan administration is probably a better prospect than the last administration, but I stand to be corrected.

As far as free riders are concerned, Mexico and a lot of others are all free riders under Part IV of the GATT, which was solemnly agreed on by the signatories in the mid-1960s. The developing countries were not required to offer reciprocity. It was authorized that they be free riders.

Thank you, Mr. Chairman.

Chairman Streeten: Ed Hamilton?

Ed Hamilton: I have a brief response to what Gerry Helleiner just said about the current United States administration. It is on the one hand correct that as an ideological proposition — and whether ideology continues to mean anything is debatable — this administration is probably inclined more toward market forces and less toward intervention than many of its predecessors.

The other side of that coin is that it's probably less inclined to pursue free trade alone than any of its predecessors. Therefore, if one can make the case effectively that the United States is following a more liberal regime than its trading partners, and if one can make that case sector by sector, one will find — at least in the early years — substantially less devotion to the liberal trade-oriented institutions and the principles reflected by these institutions than one has seen before.

So the interesting issue for this administration is likely to be the "bottom line" in each major trading category, and the balance of actual interests and forms of behavior: for example, nontariff barriers that are, in fact, making a mockery in some item of the international free-trading principle. The United States is not going to be as likely to say, "Well, the international free-trading principle, after all, is a form of our religion, and it's expressed by a particular institution called the GATT, and, therefore, we should overlook this short-term transitory situation with this or that trading partner in this or that item, and think of the overall objective."

Therefore, if one takes comfort in the ideology of free trade, one has to recognize that there will be a very severe attitude toward people who don't reciprocate.

Chairman Streeten: Mahbub?

Mahbub ul Haq: I have been asked to respond to two questions from the audience. The first one is, "Would someone rebut Walt Rostow's outcry over why the rich countries should be expected to surrender sovereignty to a South that is unwilling to surrender any of its own sovereignty."

I won't take advantage of Professor Rostow's absence at this stage, but I know that Professor Rostow is a historian and certainly not unfamiliar with the South's surrender of sovereignty for a very long time, politically and also economically. But to be serious, I think that talk about surrender of sovereignty is based on the wrong premises.

Let us look at the real world. The South lives largely in an economic world fashioned by the rich nations. The monetary and credit system, the resource transfer system, the foreign investment system, the technology transfer system, the research and development system — all are largely in the hands of the developed countries, which have a far greater degree of control over these systems than do developing countries. And the South is not unfamiliar with policy conditionality imposed by the IMF and the World Bank, which require fairly drastic changes in domestic economic policies.

So I don't think it's a question of whether the South is unwilling to surrender a certain amount of sovereignty. I think "surrender" is the wrong term. The right term is "search for partnership" in running the world's economic system. If we are moving into a more interdependent world and, as Gerry Helleiner was saying earlier, if the potential of the markets of the developing countries is grossly underestimated today, and if in the next 50 years 90 percent of the world's population is concentrated in these countries, the question is: What kind of international institutions will be required? They must be institutions of interdependence in which there is no domination but rather a sharing of power, a partnership of mutual dependence.

A partnership involves limiting some of the power that the rich nations have enjoyed. I won't quite call it surrender. This is painful. And I think basically all nations are beginning to recognize the limits of power, political and economic, and the question is how well one devises a world with institutions that balance various power groups and establish these principles of partnership. That is the real challenge, rather than a surrender of one side to another.

And this links up with the second question that has been posed to me: "Could you please detail how the resources for the World Bank can be internationalized? This refers to your comment that the composition of the shareholders will alter the institution."

Well, let me first describe my vision for the late 1980s and 1990s. I believe that the current monetary and financial institutions, or the Bretton Woods institutions, are an endangered species from both sides, from the left and the right, from the creditor's side and the borrower's side. I think that there will have to be a search for new institutions to manage world interdependence, and I see three institutions emerging by the end of this decade.

First, I foresee an international central bank which is a culmination of reforms Paul discussed: some institution that manages the creation and distribution of international currency. Incidentally, those who talk about surrender of sovereignty should recognize that the United States has been exercising this sovereign right on behalf of all of the nations for the last few decades, and most nations have very little say regarding the quantum of international liquidity or its distribution.

But I think there is going to be a major conflict on this issue, and ultimately if international currency is to be created, managed, and distributed, we need an international central bank. We came to that conclusion within individual states after a good deal of struggle. We'll come to it at the international level, I predict, by the end of this decade and have an international central bank.

If we have an international central bank, it also becomes possible to link up some of the international reserve creation with development needs, because then the bank can have two windows: one for short-term balance-of-payments problems and another for longer-term development. This was the original Keynes conception anyway. The distinction between short-run and long-run finance is rather a thin one in this case.

The second institution emerging as part of this will be a world development financial fund, partially from the creation of international currency and partially form international taxation.

I do believe that automaticity in the creation of international money and international taxation are around the corner. This will sound very strange in a period when countries are unwilling to tax themselves for even their own domestic needs, but I think there are certain mechanisms which will begin to perform these functions. For example, OPEC could well become, and already is, a taxation authority for the world. In a recent meeting one of the OPEC members, Kuwait finance minister Alhamud, proposed a certain tax on petroleum exports which might be credited to an international resource transfer institution. Even a tax of $1 per barrel would yield about $15 billion a year.

But however it comes, I see the internationalization of resources and a world development fund of the kind that the Brandt Commission spoke of in its report.

A third institution, the International Trading Organization (ITO), is implicit in what Gerry Helleiner said. If we are to manage the trading system with more participation by both sides, then the GATT is, again, an endangered species. There will be pressure for something which combines UNCTAD and GATT functions and sets up an international trading organization along the lines that were originally conceived in the mid-1940s.

I frankly have not the slightest idea of how these international institutions of interdependence are going to emerge, just as no one could have predicted in 1940 that World War II would conclude with the burst of innovation and new institutions to manage the world.

But we are in a period of major crisis and tremendous confusion about the nature of appropriate policies. I don't get discouraged by that because I always feel — as I am also a student of history — that in those times of confusion and crisis comes the great burst of intellectual activity and innovation. And I believe the 1980s will see creativity of the kind that probably we have not seen since the 1940s. This may sound more like a hope and a prayer than established fact, but I will go by that.

Chairman Streeten: Nicolas Barletta?

Nicolas Barletta: After the wide optimistic vision of Mahbub, I don't want to sound overly pedestrian, but I have a couple of comments about some of the things that have been said. I want to stress their importance for the solution of concrete problems.

Of great significance are the new policies that the IMF itself has been adopting, by which the adjustment processes of borrowing countries would take place over a longer period of time. I think there is immense significance to this new instrument of action, which is still being tested imperfectly. It gives the developing countries the opportunity to stretch out the repayment of borrowings necessary for the adjustment processes that they have to adopt in order to stabilize and go back to a trajectory of growth. The financial and balance-of-payments problems are getting them into political trouble with the very stern one-year and one-and-a-half-year conditionality factors, because the adjustments needed in such a short time are too abrupt and socially painful. This new IMF instrument becomes extraordinarily important in the present period, when most of the balance-of-payments problems being faced by the developing world are not of their own making, but the making of the developed world and the OPEC countries.

I agree with most of what Mahbub said with respect to the wide evolution within the World Bank, not in all the details, but I will not take this opportunity to air our disagreements. There are a lot of ways in which both of these institutions can extend and expand their usefulness to the developing world in the next few years in very important matters.

With respect to the monetary system, I tend to agree that the flexible exchange rates mechanism of the last few years has proven to be extraordinarily good for the world. If we had not had that flexibility in the exchange markets, I think we would have had a major collapse of the whole world economy, perhaps even worse than the one we had in the 1930s. And this is something that we should also underline: not only the bad things that have happened, but the worse things that have been avoided.

However, I think we also have to take into account the fact that flexible exchange rates which have these overall beneficial effects can have some ill effects by creating asymmetry between the strong currencies of the world and the weaker currencies of the world. When the adjustments are made, let's say, in the dollar market or in the Deutsche mark market or in the yen market, we also need ways of minimizing the effects of these adjustments on the currencies and economics of developing countries. But that is really a way of tinkering and correcting, not an overall change in this new mechanism of flexibility that has been introduced.

Chairman Streeten: Thank you. Roger?

Roger Hansen: Thank you, Mr. Chairman. I hope I don't stray too far from the very interesting discussion that we have had thus far on the issues that are specifically before us this morning.

The two problems in my mind, I suppose, arise because we are drawing closer to our final session, which asks whether a mutual agenda can evolve that will allow North and South to move beyond the present highly constrained conflictual situation, at least at the rhetorical level.

And if we are thinking about where we stand at the end of these two days of discussion, then one cannot help reflecting for just a moment on the exchange between Professor Helleiner and Professor Weintraub, not because either is right or wrong, but because that exchange captures what strikes me as the vital essence of the problem to which we are seeking an answer. Presumably the heads of state and others who go to Cancún will be seeking an answer to the same problem.

So using Helleiner and Weintraub and their interchange as the evidence, let me identify the two fundamental problems I see in the North-South debate at the present time, and the difficulty of moving beyond the present stage of the conflict or stalemate.

The first problem involves the conflict between the view that economic progress, in the developing countries in particular, has been good, indeed very good by historical standards, and will continue to be good even if at slightly lower rates of GNP growth, and the view that we may in the 1980s be facing just enough discontinuities, economic or political, to make the 1960s and 1970s irrelevant to the problems of the 1980s.

And one of the difficulties here is that there is evidence to support either point of view. Somebody who believes that the progress of the past will continue can cite growing degrees of interdependence that cause the Trilateral Commission, multinational corporations, and the international financial community to have a great stake in seeing that the 1980s do not lead to a collapse of either the GATT, in terms of trading rules, or the IMF, in terms of recycling capacity, debt relief problems, et cetera.

In other words, there are already so many Northern interests involved in preventing the total collapse scenario that one can view the 1980s as a period in which there might be some rough seas but no serious backsliding. From this perspective, one can feel rather comfortable with the 1980s, despite the present difficulties.

On the other hand, those who regard the 1980s as qualitatively different and the record of the 1960s and 1970s as not that relevant may be thinking of different and novel issues and contexts. That is, they are thinking about the potential impact of the energy problem in the 1980s. They may be thinking of the potential problem of food production and distribution in the 1980s as well as environmental issues and population issues. If you put these foreseeable and fairly predictable trends together, it is quite feasible to argue that the degree of international management necessary in the coming decade will require a quantum jump. And if this the case, then one can indeed argue, as Gerry Helleiner seemed to be arguing, that there is something close to irrelevant about extrapolations from the past.

So again, we are left with the problem in which both sides can, on the basis of selective evidence, feel very firmly that they are right and the other side is wrong. And this problem is faced every time the

North-South issue is raised. In sum, a data base can be developed to support widely differing perspectives on the North-South problem.

The second problem involves what I perhaps have inappropriately chosen to call the "legitimacy" issue.

In the Helleiner and Weintraub exchange, this issue was certainly implicit, if not explicit. The Helleiner view, which came to the surface at the very end of his remarks and then in his response, was that regardless of the past, there should be some changes made in the system because there is something presently lacking in terms of legitimacy — whether we think of that concept in terms of the present rules of the institutions we have been discussing this morning, or whether we choose to define legitimacy mostly in terms of equity, justice, fairness, or any other similar concept. The Helleiner view was that there are so many biases and imperfections in the present systemic norms and rules that it is not only fair but inevitable that the developing countries will question them.

I take it from Helleiner's remarks, and again from the North-South debate, that the calls for "structural reform" on the part of the South have little, if anything, to do with rates of growth in the 1960s and the 1970s. They have to do with a view concerning a perceived inequity involved in the present set of rules.

I'm not saying that their view is correct, or even that the issue should be posed, because posing questions of equity often creates not only tremendous conflict and deadlock, but often perverse results. But then, Americans like to duck the equity issue, and it doesn't surprise me that we would like to duck it internationally as well as domestically for what some consider a very good reason: that once raised, the issue leads to more conflict than compromise.

But if the Southern view sees this legitimacy issue as of fundamental importance regardless of past growth rates and therefore demands alterations, the Weintraub view implicitly accepts the legitimacy, fairness, and equity of the present system. According to this perspective, we live in a world in which nothing is perfectly equitable or legitimate. We live in a second-best or third-best world, and in that sense, what we have is about the best we can expect.

And this, indeed, was Sid's view in pointing out that growth rates for less developed countries under the present system have been better than one ever would have predicted. Indeed, these rates are better than most economists predicted at the beginning of each decade when they set growth targets.

So Northerners either accept the "legitimacy" of the present system specifically, or from a pragmatic point of view they dismiss the issue by saying that legitimacy is not or should not be allowed to become an issue in the North-South debate. They feel it will be more detrimental than constructive to a resolution to the problem. The Southern response to this viewpoint is that it dismisses as incidental or, in a sense, nonnegotiable what is for the South the essence of the negotiation. The problem is a fundamental issue of legitimacy.

In terms of the negotiation process itself, what matters is not whether there *should be* an issue of legitimacy, but whether or not there *is* one. It may be inappropriately brought to the table in the views of some people, who say it will confuse and confound the issue, but if it is there and is going to be brought up by eloquent participants in that negotiating process, then we have to deal with it. That brings me back to the Weintraub exchange with Helleiner. That kind of exchange, it strikes me, captures in one sense the essence of seven and a half years of one side talking to the other without making any progress. The perspectives are so different that it is hard to see what a mutually accommodative negotiating agenda would be.

Chairman Streeten: Yes, Mahbub?

Mahbub ul Haq: Well, I think Roger raises a very fundamental issue. What is it that is bothering the developing countries if the growth rates are fine and if they can get a lot of money from the World Bank and the IMF? Why do they bother so much about who sits on the Board? Why do they bother so much about who runs the institutions, so long as the end result is okay?

First, we have to differentiate the developing countries. I think the developed world, which lectures us so much about differentiating the Third World, forgets these distinctions when they talk about growth prospects. Growth prospects are very different for the middle-income countries and the poorest countries. The current projections of the World Bank are that by the year 1990, middle-income countries will reach per capita GNP levels of $2,000, and poor countries, $200. These are poor countries with per capita income now below $350, but they are the majority of the Third World population. So if you look both at the growth record in the 1960s and the 1970s and the growth prospects, one of the things that comes through very forcefully is the widening gap between the middle-income countries and the poorest countries. This gap is widening at a much greater rate than even the gap between the entire Third World and the developed countries. So the first issue is what the world structures have to do for the poorest countries.

Second, just looking at growth rates ignores the question of frustrated economic growth and frustrated opportunities. Is an average growth rate in the next decade of 5 or 6 percent enough for the developing countries? It depends on the perspective you start from. Just because they have to make up so much in such a short time, there is no reason why they shouldn't shoot for 10 percent or 15 percent growth rates, like Japan and other countries did at certain stages. It looks very ambitious in the current context, but they are trying to push the frontier of opportunities in this.

And third, of course, I agree with Roger that even if one could achieve maximum growth rates and all the resource transfers required, the issue of legitimacy of international institutions and rules will remain, along with what kind of institutions should manage interdependence and what kind of partnership there should be in the future. These questions ae very vital to the policy frameworks which will emerge internationally and nationally.

Chairman Streeten: John Sewell.

John Sewell: I wanted to raise an issue which is particularly ticklish, I suppose, but I raise it in light of the fact that a group such as ours, hopefully, can do more than replicate the official discussions and disagreements that go on.

It seems to me that one of the great problems we face is that we have gone through a discontinuity in the world's economic structure and its prospects. But at the same time we are still in the midst of a long, continuing process of accretion of power to nation-states and the rise of nationalism, which is very real both in the North and in the South. I'm not sure whether the North is particularly prepared to give up some of the major attributes and powers of the nation-state that may be necessary for dealing with the problems we have been discussing or for enhancing the legitimacy of international institutions. And I'm afraid the same thing is true of nation-states in the South. These countries have fought long and hard in many cases for political independence and for sovereignty, and the pressures to maintain their status and not to give it up will be very great.

It was only after Europe when through a series of disastrous wars and depressions, brought about by the inability to delegate to some degree national sovereignty, that it was willing to create the framework of the European Community.

Neglecting this strong impetus within the developing world and within the North may lead to a certain degree of unrealism in our discussions. I would sense that at least part of the reaction of a country like Mexico vis-à-vis membership in the GATT has its origins in a sense of national sovereignty.

The limited record of cooperation among developing countries in the South and an inability to transcend national sovereignty is not taken as a reflection of fault, by the way — just as a reflection of the real world.

It is often said that there could be a TEAPEC, an OPEC for tea, if only India, Ceylon, and Kenya could agree. And you really wouldn't have to have an OPEC; just a slight increase in the price of tea per pound would bring great benefits to all three of those very poor countries.

There has been a great deal of discussion and analysis concerning the need for the equivalent of the OECD for the South, with some sort of analytical capacity to deal with public policy issues. It could be started tomorrow if anyone would put up the money, and there is a great deal of money in some parts of the South that could fund that kind of operation. But it hasn't happened.

Part of the objective is not only a more equitable and a more efficient nation-state system, but also the alleviation or the amelioration of the condition of some very poor people around the world.

The barrier of the nation-state concept is going to be very great, because the nation-states are too small for the big problems and too big for the small problems. They don't deal with local problems very well, either.

When one talks about the legitimacy of the system, one implicitly talks about giving up some degree of national sovereignty, which is going to be very difficult in the troubled 1980s and 1990s.

Chairman Streeten: Joan Spero.

Joan Spero: I want to go back to something that I raised yesterday and that Mahbub raised a moment ago: that our greatest hope is an international crisis. I have one in mind. Our greatest salvation might well be an international banking crisis. I wanted to raise in that context the whole problem which we have touched on in bits and pieces but which no one has addressed directly.

In my view, one of the major changes that has taken place over the past decade and a half in North-South relations, let alone in Northern and international relations, has been the whole change in international financial relationships. And if we apply it particularly to relations among developed and developing countries, it's been the huge surge in lending from private banks to developing countries — admittedly to a relatively small group of developing countries. The Bretton Woods system and the Bretton Woods institutions have really changed the whole structure of the international system, from the internationalization of banking to the creation of Euromarkets and the access of certain developing countries to those markets. Despite certain arguments that we're reaching the limits of lending and that we can no longer increase at the previous rate, I see important signs that lending will increase in the 1980s. I don't see banks cutting back their lending. They are still acquiring deposits, from oil-producing countries primarily, which they will continue to lend to developing countries.

In a reformed Bretton Woods, we need to think about a way to keep commercial flows going to developing countries. This can be done in a variety of ways, but it seems to me that the process has a lot of implications for the Bretton Woods system.

One thing to do is to increase the way the private system and the Bank and the Fund work together in order to continue flows. Cofinancing is one way. Cooperation between the Fund and the private banks in debt rescheduling is another. One other possibility is the access of the private banks to IMF information and analyses. I realize that this is a very touchy issue, but it is one to consider if we are thinking about keeping the flows going.

Another dimension of that issue has more explicitly to do with the banks: the possibility of some international lender of last-resort capability. It could be an IMF safety net for banks or perhaps some kind of international insurance scheme for lending to Third World countries.

There is also the whole issue of capability for debt restructuring and debt renegotiation. We need to develop more regularized procedures, although they are gradually emerging, for debt renegotiation, which would articulate the public interest as well as the private negotiations. We're moving in that direction through the Paris Club, and UNCTAD is putting a hand to these things. The Fund is also working at them. Here is where we need not a new institution, but a new coordination of institutions.

And now I come back to Gerry Helleiner's point, because I think he is absolutely right. This is all related to the GATT. If the borrowing countries are going to be able to repay their debts, they're going to have to have access to Northern markets. We're going to have to think of some way to ensure market access, whether it's in the debt renegotiation process or the GATT process.

Let me conclude with one comment, although it leaves out the less developed countries because they, at this point, do not have access to the markets. But if you can somehow develop more structured relationships between the Bank and the private sector, then maybe you can open new markets or open the existing markets to more countries.

Chairman Streeten: Thank you. That was a very important area which we have neglected, I suppose, for reasons of time, but I am extremely grateful you covered it. Don Mills?

Donald Mills: I have had the great privilege, if not the distinction, of being at the Fund for a couple of years as a humble alternate executive director. So I am ambivalent about that institution. I have great respect for it, but I also believe that I have seen its faults: the way that it doesn't really focus on the fundamental circumstances of developing countries.

The Fund and the Bank are in a way like Jack Sprat and his wife: "Jack Sprat could eat no fat, his wife could eat no lean, and so between them both . . ."

Now the trouble is that the Bank has been nibbling a bit of the Fund's meal. The Fund had been very slow in moving on the matter of providing an adequate period of adjustment to countries which use its resources in times of difficulty. The Bank has moved into this territory by providing structural adjustment loans.

There are many differences between these two institutions and one must accept this. But my impression is that the Bank has adjusted more successfully and more rapidly to certain realities than the Fund. I don't want to be unfair to the Fund, but that is my impression. I wonder what it is that makes the Bank more capable of adjusting to realities. Is it leadership? Is it the fact that the success of the Bank depends on the success of its projects? With the Fund, if a country fails, it is seen as the country's fault. If the developing country mismanages its economy, the Fund is right and the country is wrong. Perhaps the Fund's pride is not as involved in a country's success as the Bank's pride is. Also, the Fund has not in the past seen development as a part of its real concern.

But if one is going to influence these institutions, one should have an idea of what makes them move and what doesn't make them move.

It is interesting that the OECD countries would now be concerned about the growing internationalism of the Bank, particularly when we recall that the World Bank, or to call it by its full name, the International Bank for Reconstruction and Development, was set up for the reconstruction and development of Europe.

Part of the difficulty in discussing the Third World with some people in the North is that they have the feeling that most of the resources for the development of the South are coming from elsewhere. And without denying the value of outside resources, this makes intelligent analysis or discussion almost impossible.

I add to that the grave misconceptions about development aid and, of course, the question of equity raised by Roger Hansen. In my experience the North-South issue is seen by many people in some countries as a question of aid. When you raise the issue, they say, "But how can the taxpayer or the voter be asked to give more to the developing countries?"

And when you try to raise the issue of equity, the issue of the legitimacy of the system, the issue of the imbalances in the system and the disadvantages facing the Third World in such areas as trade, you have a hard time. These issues are not discussed in the media, particularly in this country, although they are discussed to some extent in some other countries of the North. It makes it very difficult for the public in a country like the United States to come to grips with the issue of equity or "fairness."

In the Federal Republic of Germany there was a great deal of misconception and superstition about overseas development assistance, and a feeling that aid was going into the pockets of a few rich people in poor countries. So the German government some time ago launched a publicity campaign to disprove the nonsense. Now, it doesn't mean that there are not any inefficiencies and even a bit of venal activity here and there. But to get rid of the notion that the money was being wasted and that it was going down the drain, the government went to the trouble — to its great credit — to launch a serious publicity campaign.

More recently, the United Kingdom government conducted a survey to find out what people thought about overseas development assistance. The first thing the people thought was that the British government was spending more on these poor countries than it was spending on its own people in Britain in some vital sectors. Then the government gave the respondents in the survey factual information and repeated the original question. This time they found totally different attitudes.

There is a prevalent notion that the poor of the rich countries are subsidizing the rich of the poor countries. My response has always been that the poor of the poor countries have been subsidizing the rich of the rich countries for a very long time, indeed. That doesn't mean that we must not strive for economic efficiency and for equity in our own communities. But it is this sort of perspective that we must overcome if we are going to have a decent discussion.

We are accused of being inefficient or being venal in the matter of development assistance resources. Yet public reports have shown the very great difficulty of managing poverty programs and other welfare schemes in the United States or elsewhere in the North and of preventing significant funds from going astray and into the wrong pockets. The moral, of course, is that it's not easy to deal with programs of assistance, whether they are on a national or international scale. And if we got together and tried to iron out the difficulties to make sure that such programs do reach the people who need them most, then I think we'd get somewhere. But we tend to get hung up on misbeliefs which do not help in the whole process of trying to establish a dialogue.

What Mahbub says about the World Bank's role in professional clarification of global issues is, in my view, absolutely right. I think that in some matters the Bank could stand between the people who are trying to negotiate for the North and the South and provide some helpful analysis on some of the issues on which we disagree.

I agree very much with what Paul Streeten says about institutional lag, and it also applies to the South in terms of its failure so far to establish some sort of secretariat for the Group of 77.

Finally, regarding the very interesting discussion between Professor Weintraub and Professor Helleiner, I would not dare to rush in. But I would say this: I acknowledge the great growth of the economies of some countries of the South in the 1960s; but during that period the gap between the developing and developed countries widened nevertheless. And what is more, one's view of progress depends very much on whether you measure it from the bottom or from the top.

Chairman Streeten: Nicolas Barletta.

Nícolas Barletta: The subject is fascinating, and there are many different viewpoints. I want to highlight two or three of them.

First, in the area of trade, I share some of Sidney Weintraub's views: the capacity for expanded exports by the developing countries should not be underestimated, even under the present system. Obviously, that does not mean that we don't need to continue improving the system, as he has said, but it's very important to bring the two things in focus.

Many of the medium-income countries — not only the Asian-Pacific countries but also Latin American countries like Brazil, Chile, Peru, and Uruguay, just to name a few — have had the capacity to increase their exports by at least 5 percent per year in real terms over the last five or six years.

So it is important that we bring a balanced perspective to these negotiations, especially when they become acrimonious. Both sides criticize each other, but they are not willing to recognize what each has to do in order to correct the problems. And it is very important that we, the developing countries, recognize that we need to do more things for ourselves, and that we already have the mechanisms and the potential to do more things for ourselves. At the same time, that doesn't mean that the North should not continue helping with many other things that need to be changed in order to improve the system.

My second point concerns Joan's comments about developing the private international capital market. Mahbub said something before which is true: that the United States expanded world liquidity and benefits from it.

During the late 1960s and early 1970s, on the other hand, we cannot say that it was not beneficial to the rest of the world up to a point. There was a cost to it, and I'm going to come back to that in a minute. But the benefit has been precisely the development of a private capital market in the international arena, which had not existed since the last recession, the world depression of the 1930s. The international capital market in the private area had pretty much collapsed. The only vehicle was private investment.

On the other hand, that excess United States liquidity, which spawned the Euro-dollar market, has managed to continue transferring a tremendous amount of resources to the developing world, especially to the middle-income world. It has also managed to make recycling easier after the shock of adjustments that took place with the oil price changes.

And now, in a sense, we are coming back to an area that would seem to be the logical one. The private banks have committed excesses themselves. Sometimes they have overlent to countries. Sometimes the countries have not known how to use the money best, and then they find themselves facing adjustment problems. Then the IMF and the World Bank are called on to help improve the situation.

Now we are examining the possibilities of how the private banks, the Bretton Woods institutions, and the countries themselves can work a little better together. This may be one of the key developments of the 1980s. Sometimes it takes a major crisis to really create new institutions, but let's not forget the price that generally accompanies such crises. World War II was a fantastically high price for the world to pay for some institutions that have been helpful after World War II. Major crises are not necessarily to be welcomed. If we can continue altering what we already have at hand, we may well minimize the cost, even though the benefits may be a little longer in coming.

A final comment concerns the interesting issue about inflation as it relates to the capital markets. Part of inflation obviously has been created by the excess liquidity generated by several countries, especially the United States, on the one hand, and by the OPEC actions with respect to petroleum on the other. The OPEC countries have tried to create a massive transfer of real wealth from the rich countries to themselves. But, of course, this also affected the developing world.

Since nobody is yet willing to decide who is going to pay the price of that inflation, who is going to minimize the loss of real income, then inflation continues to manifest itself.

However, what is the cost? The OECD countries are paying the main cost in inflation. What we observe in this country is that the American people are tired of inflation. Part of what they voted for the last time around — whether it will happen again or not is something else — showed that they were tired of inflation. That is the cost being paid mainly by the OECD countries.

The medium-income countries are indebting themselves, and we all know that the best situation in inflation is to have debts, because inflation will erode the real value of debt. So one benefit that the medium-income countries are receiving right now through their huge indebtedness is that the debt is being eroded with the inflation that has taken place. On the other hand, the cost that they are paying is a cost to growth and employment generation in a period in which they need to sustain growth and employment generation in order to maintain political stability.

The real tragedy and the thing that we need to concentrate on is that inflationary pressure is harming the very poor countries of the world the most. Just as within domestic economies, it tends to hurt the very poor people most. This is something we cannot forget.

Chairman Streeten: Jorge Navarrete?

Jorge Navarrete: I will comment just on a very specific point: Mexico's position in not joining the GATT.

Mexico decided last year not to join the GATT. The decision to postpone this possibility was taken after a long national debate in which different national and international factors were taken into account. The main basis for the decision was, in my view, the following.

First, it was felt that Mexico needs full control of its internal development policy, and that includes control over foreign trade flows. There is great concern over the necessity to control foreign trade flows because of several features of Mexico's economic growth history and its present economic structure, which is heavily dependent on imports. The elasticity of imports to growth is very high, particularly in the sectors of capital goods and intermediate goods. For this reason, Mexico's development processes as a whole are very much linked with foreign trade factors.

On the other hand, the prevailing internal distribution of income favors nonessential imports, and there is a need to control these luxury imports if a rational use of foreign exchange receipts is to be achieved.

Given the recent growth of oil exports, Mexico's general import capacity is increasing, and it is essential to control and channel this growth in order to avoid the kind of unfavorable situations that have arisen in other oil-exporting developing countries.

But despite all this, the important factor is that Mexico is perhaps now the fastest growing import market among developing countries, and this rapid growth of imports will continue in the foreseeable future at very significant rates. So Mexico is and will be an open market in many senses, but is trying and will continue to try to channel an increasing percentage of imports to those goods that make the most positive contribution to the development process.

Of course, the GATT decision also took into consideration international factors. The GATT is an organization more devoted to promoting and organizing trade among developed countries than in a truly worldwide sense. I think recent developments confirm this general orientation.

I agree with the picture that Professor Helleiner offered on the new protectionism in the 1980s. There is a need to approach this question and try to resolve not only the short-term questions, but also the long-term perspective of industrial development in developing countries and of industrial restructuring in the developed world.

Gerald Helleiner

Chairman Streeten: Gerry?

Gerald Helleiner: On the IMF and its functioning, it seems to me that there is one set of issues that we have not addressed, but which is extremely important and not generally recognized. These issues follow in part from what Joan Spero said.

The function of the IMF, after all, was to provide capacity to deal with short-term balance-of-payments shocks. Over the course of the last half-dozen years, a whole variety of changes has emerged which have affected individual countries' capacities to deal with short-term balance-of-payments shocks. One change has been this enormous expansion in commercial bank activity, which has been of great consequence to a selected number of countries. The banks have replaced the IMF as a source of short-term balance-of-payments assistance, but only for a dozen or so developing countries which, I have no doubt, will continue to have reasonable access to the private banking system with the odd hiccup here and there for particular countries.

The new debt situation has some major disadvantages, however, even for these countries. Inflation is, of course, eroding the value of the debts. On the other hand, interest rates are now very high in anticipation of future inflation — and now fluctuate, often being revised every six months — so that a country like Brazil, with 60 percent of its exports devoted to debt servicing this month, can no longer know whether this figure won't be 70 percent next month. This is a very dangerous, very uncertain, and very skittish way of managing the international balance of payments. Still, there it is. It's new and it's extremely important for some.

Second, there has been an enormous increase in the price of gold. This affects the capacity to deal with short-term balance-of-payments problems, for the major recent source of increased reserves, in the traditional sense, has been gold price expansion. Three-quarters of the world's gold is held by seven countries: the United States, the United Kingdom, Switzerland, France, Belgium, and Holland. It's not just a matter of the 100 million ounces held in the IMF. It's the many more million ounces held by these nation-states. The developing countries are not among them.

In addition, there are all sorts of new agreements among the developed countries and among banks themselves. The European monetary system, which in fact has monetized a lot of gold already, provides further credit for short-term balance-of-payments needs.

And all of these arrangements are virtually automatic and unconditional.

Now the question is: what happened to the IMF unconditional lending and the SDR system?

The SDR was supposed to be the basis for expanding world liquidity and the capacity of countries to deal with short-term balance-of-payments shocks. It is now the view of developing countries — whatever it was in the past, Sidney — that SDRs should resume their planned role as the center of the international monetary system.

SDRs, in fact, accounted for about 1 or 2 percent of the recent enormous expansion in nominal world liquidity, not even counting access to commercial bank lending.

Those people who got stung on all this, as usual, were the weakest and smallest of the developing countries, who neither had gold — in fact, the United States Treasury persuaded them for years not to convert their dollars into gold, and then after the revision of the Articles of Agreement of the IMF, they were forbidden by law to convert them into gold — nor access to the commercial banking system. And now the IMF comes along and tells them that if they need to borrow, they have got to impose some very restrictive conditions. These conditions have been adjusted, as several speakers have noted, but have not been adjusted enough. Furthermore, the present debt problems have been created externally, *not* domestically. The low conditionality, automatic access to balance-of-payments assistance which is conditionality, automatic access to balance-of-payments assistance which is available for virtually everybody else is not available for the bottom thirty or forty countries. Yet they are the ones with the greatest instability of export earnings, the greatest rigidity in their economies, and therefore the greatest cost of making rapid adjustments to serious balance-of-payments shocks. There is no greater inequity than that which has emerged in the distribution of global liquidity in the 1970s.

And we don't need an SDR-aid link to restore a degree of sanity and a degree of order to the system. To restore the system a little bit to what was originally planned, all one has to do is to eliminate the view that we must not create SDRs, because the world is so full of liquidity. The argument is absurd. The SDR has not been a major source of liquidity expansion for anybody except these poor thirty or forty countries at the bottom.

What is needed now is the IMF's recognition that it must expand substantially, more than it already has, low conditionality access for those poorest countries who are experiencing difficulties through no fault of their own. Or, if this is too difficult, what would perhaps be much more likely, much more hopeful, and very reasonable is to begin on January 1, 1982, with the creation of substantially more SDRs than we have had in the past. The rationale for the creation of SDRs is and always has been the international level of global need.

"Global need" now means the provision of short-term balance-of-payments assistance for those who do not have alternative sources of this assistance. You can work out the figures without introducing the SDR-aid link. The developing country group has about $60 billion worth of reserves. To provide them with a 10 percent expansion per year (which works out to about $6 billion) requires, under the existing formula, the creation of about $25 billion of SDRs every year. That's not an exorbitant amount. In planning for the 1980s I would say it's extremely conservative.

We need something on that order to restore a degree of equity in the distribution of capacity to deal with short-term balance-of-payments problems and, in particular, to return the world to the system that we thought we were planning.

Chairman Streeten: Well said, Gerry. Thank you.

In my summing up remarks on this session I first want to reply to one of the questions from the floor: How should we act in an uncertain world? Yesterday John Foster talked about the uncertain future of oil, but of course this uncertainty applies to other crucial items. We don't know what the situation will be in the next ten years, and many projections, whether with respect to oil or steel or shipbuilding or fertilizers, have turned out to be wildly wrong.

The discussion yesterday left us with several questions: What are we to do? How do we plan? How do we behave in a situation where the future is uncertain?

Planning in the face of uncertainty raises important issues which I can't resolve now, but I just want to tell you one of my favorite analogies. It applies to energy, it applies to steel, it applies to anything that is crucial for not only growth but also equity and distribution and the objectives we want to pursue. It is the story of the Swiss hero, Wilhelm Tell.

If you remember, he had to shoot an apple off the head of his son. I'm not sure whether anybody sensible was there to give him advice, but a good economist would have said, "Aim high." If he aims correctly, then he shoots the apple off. If he shoots low, he will kill his son. But if he aims high, he at least avoids killing his son.

If we don't know whether there will be a surplus of oil in 1990 or

2000, or whether there will be a great scarcity or deficiency of energy, it seems that a sensible policy to pursue, if we regard energy as an important input for growth, is to aim high. The risks of having a surplus are less than the risks of being caught short and having to hold back world growth.

Now, let me turn very briefly to a number of points which have emerged from this morning's conversation.

First, the discussion about international institutions brought out very clearly how we all judge the progress of these institutions differently. We either regard them as a kind of still photograph — comparing them to our ideals of how we would like them to work and finding them very deficient — or we judge them as a film, as a kind of evolution of responses to changing world circumstances, comparing progress over time with the past.

In some ways the Fund has been more adaptive than the Bank to the new challenges, considering all the new financial facilities which the Fund has created.

But at the same time, the world has changed even faster and, therefore, in relation to what we would like it to be, the gap between needs and responses has increased. So how do we judge these institutions? Do we judge them in relation to adaptation to new innovation? If we look at the Fund, I think it has been innovative. Or do we look at them in relation to what is needed now, what we would like them to do? Then the deficiencies are great.

From this perspective, the gap has widened. It's like the dog in a different corner of the field following the man. He sees the man walking and trots off in that direction, but he hasn't enough intelligence to predict the way he is going, and therefore the gap between man and dog may well increase.

This is one set of reasons why we may so often and vigorously disagree on the adequacy or inadequacy of some of the major institutions as they relate to the North-South problem.

Second, I suppose there was too much agreement among us. There has been too much agreement about the need to channel more resources, whether they be short-term, medium-term, long-term; whether they be grants, loans on concessional terms, or loans on commercial terms. By and large, most of the people around this table would agree that more resources transferred to developing countries, particularly the poorest, are a good thing.

Yet, there is mounting opinion, not represented around this table, both on the right and on the left, that the transfer of resources to the Third World is counterproductive and encourages the wrong kind of political regimes and the wrong kind of policies. People like Peter Bauer say it encourages central planning and discourages the private market. Left-wing radical critics say that it perpetuates conservative, feudal, reactionary regimes, and that it prevents the enterprising self-help and the savings that would be realized if these countries did more for themselves.

This view has not been represented, and I don't share it. I think it is faulty on logical, economic, and historical grounds. But it is a view that is gaining strength, and I think one should just mention it, because it casts some doubt on the value of the things that many of us have been advocating.

Those radical colleagues of mine, who think transferring resources to, say, Latin American countries weakens their will to save, ought to rejoice over the heavy debt service that these countries are running, because if a country has to pay out a lot of money, that should, by their argument, encourage domestic savings and strengthen domestic development efforts.

Yet, I haven't seen many of these radical colleagues welcoming the heavy debt burden, the "reverse resource transfer," that these countries have to pay.

We could talk at length about free trade. I am not a free trader, perhaps even less than Gerry. I don't know whether Professor Weintraub is one. Advocates of free trade are facing a puzzle. As Rousseau said, "Man is born free, and yet everywhere he is in chains." Most of my economics colleagues advocate free trade, yet nowhere does it exist.

I would invite Professor Weintraub and other advocates of free trade to do precisely what he suggested I should do: to analyze the reasons why free trade has not been adopted. I think there are some good arguments and many bad ones to be put forward against free trade. Even if we were very strongly advocating freer trade and if we were to judge which countries and which groups within these countries would gain from this, I suspect that the poorest countries and the poorest people within countries, which are of primary concern to many of us, would not experience any gain. Once again, it would be the middle-income countries, the Brazils and Mexicos, and the richer groups within these countries that would benefit the most.

The whole question of free trade and free access of multinationals raises the issue of the transfer of consumption patterns from North to South, and the consequent call on the part of some analysts for limited or perhaps full de-linking between developed and developing countries. This viewpoint has not been represented in this symposium. It is a perspective which advocates a de-linking of the South in order to allow developing countries to evolve indigenous development alternatives and indigenous patterns of cultural identities. I know that Mahbub can be very illuminating and stimulating on this subject, but he wasn't given a chance to do so today. This is a very important issue, and I am sorry that we didn't have time to explore it.

Let me put the question of the GATT in a nutshell as I understand it: I think the accusation concerning the free rider issue is based on a fundamental confusion between rules that are uniform and rules that are universal, but which allow for the different circumstances in which countries and peoples find themselves. In the North-South debate, people have confused uniformity with universality, and, therefore, they have argued that the developing countries want exemption from certain rules — rules about reciprocity, tariff reductions, debt repayment, loan terms, et cetera.

This involves a basic confusion. Most of us pay income tax. We don't all pay the same amount of income tax. Income tax is graduated according to needs, according to earnings and deserts, according to where our money comes from, according to the size of our family, according to whether we are blind, over the age of sixty-five, and so on. It's a *universal* system, but it is not a *uniform* system. It's a system that is adapted to the *circumstances of the individual*, but according to universal principles. Similarly, in international trade and debt and capital movements, it is not a question of asking for exemptions for one group of countries. It is a question of allowing unequals to be treated unequally, according to their position and circumstances. This principle of universality but nonuniformity is absolutely fundamental to an understanding of the demands and the claims for the international economic order.

Again, I think we could much more fully discuss the issue Mahbub very briefly and interestingly raised regarding the staffing of international organizations. I don't know whether he would agree. It's not so much that one wants better or greater representation from the developing countries. The developing countries are clearly underrepresented in the World Bank. But surely we would ultimately want to aim at an international civil service whose officials are genuinely devoted to the world community, not to their own countries of origin. Where these officials are being brought up and where they have been educated may make some difference, but the reform would be like the kind of reform that was produced in the nineteenth century for the English civil service, in which highly corrupt recruiting from particular groups was replaced with entry by merit and loyalty to the service.

We should aim at creating an independent, efficient, loyal international civil service in international organizations rather than country quotas. These civil servants would be answerable to the international community and its interests, though sensitive to social and political issues. We are still quite a long way away from that.

Finally, I'm not quite sure what Professor Weintraub intended to convey by quoting from that *New York Times* article. If it is an attack on clichés and tautologies, he is barking up the wrong tree. I am a great believer in the value of tautologies and clichés. It is useful to be reminded that the world order should be just, that changes should be fundamental, that adjustments should be structural, and that proposals should be imaginative. Tautologies of this kind prevent us from getting lost in the technicalities of means and keep objectives clear.

Let us not neglect the value of saying the obvious. A decent and just world order, achieved by innovative and imaginative proposals, is a worthy objective that is not always kept in sight in our preoccupation with the technical issues. It was James Meade who said that he would like to have written on his tombstone: "He tried to be an economist, but common sense kept breaking through." Perhaps Professor Weintraub just accuses me of not being an economist.

If, on the other hand, Professor Weintraub is saying that we ought to think hard not only about our agreed-upon objectives, which are obvious, but also about how we can implement them, how we can mobilize political and economic interests to support these objectives, I very much agree. It is a challenge to the conventional narrow look at national interests, often the result of particularly well-organized lobbies, because many interests for desirable reforms go beyond national frontiers. They unite particular groups in one country with other groups in other countries: private bankers and aid lobbies; independent retail chains, consumer advocates, and labor-intensive poor exporters; the unemployed and advocates of monetary reform. And when we talk about mutual interests this afternoon, I hope we shall return to the question of how to mobilize these interests to achieve the ridiculed "fundamental changes," "structural adjustments," and a "just world order."

Session IV
Can a "Mutual Interest" Agenda Evolve during the Global Negotiation?

Chairman Sewell: In this session we return to the crucial question with which we began yesterday morning: Is it possible to develop an agenda for negotiations between developed and developing countries based on mutual interests? If so, what would be the rough outline of that agenda, and how might the negotiations progress?

The issues central to the present North-South debate are not new and did not spring full-blown from the brow of a few Third World intellectuals after the oil price hike in the early 1970s. They go well back to the early 1960s and probably somewhat earlier. The North-South dialogue, therefore, represents one of the longest ongoing sets of international negotiations, going back before the Law of the Sea Conference, back to some extent before serious arms limitations discussions, and back almost to the beginning of post-World War II trade negotiations in the late 1940s. Therefore, they are not something which should be on the periphery of American foreign policy, as they are to this day. And they are not something which should be ignored by academic analysis, which they also have been until this day.

The pressures have continued to be very strong from the developing countries to negotiate, despite all of the stresses and strains that they have undergone since the 1970s and before, and despite a disparate set of interests within the Group of 77.

The anomaly is that everybody still seems willing to negotiate, despite the seeming lack of power of the South and the lack of interest or the preoccupations of the North. Time after time, people have returned to the negotiating table, either because they perceived benefits in doing so or costs in not doing so. I hope we will examine that anomaly this afternoon.

We are in for another year of negotiations, whether we like it or not. Around the table, someone said cynically that if the issues went away, of course, we would all be out of a job. But there will be a new round of global negotiations — a summit meeting in Cancún, Mexico, in the fall. And, of course, there are a number of other negotiations going on that are really part of the North-South dialogue, even if they aren't formally included in it, concerning trade, food, the law of the seas, and so on. The question, therefore, is really not whether to negotiate, but how we negotiate and to what ends.

Of course, these negotiations are regarded with a rather high degree of skepticism, which in the short run is probably shared by everybody at this table and most people in this room. The experience of these years of negotiations has illustrated quite clearly their difficulty, whether they take place at the United Nations, within the GATT, or elsewhere. Therefore, the question is not only whether a mutual-interest agenda can evolve during the Global Negotiation, but whether the shape of one can evolve before the negotiations begin. Unless interests are directly addressed, the prospects for the negotiations producing any concrete results are not encouraging.

Our discussions this afternoon need to consider whether it is possible to move constructively away from what everybody admits to be the present state of deadlock between the developed and developing countries.

Our first speaker this afternoon is Nicolas Barletta.

An Economic Perspective

Nicolas Barletta: My answer to the question of whether a mutual-interest agenda can evolve in the Global Negotiation, despite recent frustrations, is "Yes, definitely." And as for whether agreement can be achieved, my anser is also a positive one. Of course, the results will very much depend on how the agenda is structured, how it is presented, and how it is negotiated.

We have heard a good deal about the pitfalls in the procedures of the Global Negotiation. We have also heard a good deal about the complications and complexities involved in some of the issues that are being negotiated and about the different perspectives, not only national but also professional — which explains part of the reason why the negotiations have not gotten farther than they have.

As our chairman said, this global agenda for the New International Economic Order is nothing new. In a sense, we have been negotiating for development purposes for at least thirty years, and the agenda, with its different topics, has always been quite similar in content. We have been dealing for many years with international finance for development, with trade expansion and the opening of markets, with stability for raw materials prices, with monetary reform, with the transfer of technology to developing countries, with the development of raw materials, and with the issues relating to multinational enterprises. All these topics are parts of the agenda every time developed and developing countries get together. We have called it more recently the "North-South dialogue," but it has always been there. Obviously, the responses to the agenda topics change with new circumstances and with our knowledge and perception of how the problems on the agenda are shaping up.

I think we all will readily admit that a good deal of progress has taken place over time. Nevertheless, the development problems are still very much there. They remain very significant, and institutions, policies, and knowledge must continue to evolve if we are to cope successfully with explosive development problems.

In the financial area, we have seen a number of accomplishments: the creation of the World Bank and the International Monetary Fund; the creation of the regional banks starting in the late 1950s, such as the Inter-American Development Bank, the Asian Development Bank, and the African Development Bank; the formation within Latin America and the Western Hemisphere of an Alliance for Progress in the 1960s; the growth of world capital markets; the diversification of lending for development; and the recent movement away from financing of physical infrastructure toward financing of human development and social development. These are but some of the developments in both institutional structure and policy that have taken place in the area of finance in the last twenty-five years.

In the area of trade, significant evolution hasn't been lacking either. The creation of common markets, the creation of tariff preference schemes, the creation of the Lomé agreement, the Asian trade agreements, and certain commodity agreements are also clear examples of the evolving and permanent nature of the negotiations for development purposes.

I could also cite concrete results in the technological field, even though they have not been as abundant or dramatic as the ones in finance and in trade.

Sometimes these agreements have been reached at regional levels; others have been reached at a global level. So this has been a continuous process. What has changed is the format, in accord with new circumstances, new problems, and new perceptions. We must keep the concept of the global negotiations of the last seven years within the historical perspective of the last thirty years. The negotiations for a New International Economic Order is nothing more than a slogan given to the same ongoing process that has been taking place. They are neither at the beginning nor at the end of the process.

This process, though many times a frustrating one, has been a highly educational and informative one. It has narrowed the distance between the perceptions of different groups, and it has identified mutual interests in some key areas. So we are the richer for it. On the other hand, we have heard some participants suggest that perhaps the process has been too global, too comprehensive, and at times too ideological.

The timing of the NIEO has not been the best. We have not had a full-blown crisis, but neither has the world economy been particularly healthy. And the problems of power-sharing, which have been inherent in the negotiation process, have added to the difficulties of reaching some concrete agreements.

It's useful to keep in mind some of the fundamental economic objectives that we are trying to pursue with these negotiations.

The first objective is to sustain the process of economic growth which many people have called into question in recent years. Just look at the evident facts. Tremendous population growth is still taking place, especially in the developing world, which means a tremendous increase in the labor supply. In Latin America, for example, this growth is going to be around 3.6 percent per year for the next ten or fifteen years. This situation requires employment opportunities, and employment can only be provided with growth. Fifty percent of the population of the so-called Third World is under twenty years of age, and tremendous investments need to be made in these people — investments in health, education, nutrition, and other areas — in order to help them develop their productive potential. The tremendous rate of urbanization that has been taking place in the developing countries is just one of the dramatic reasons why growth continues to be a necessary condition for the world for the next fifty years at least.

In addition, the rapid development of expectations of all the world's people, reflecting advances in communication, education, and health, means that they are not going to await rising standards of living passively. They want to see significant improvements in their lives.

Growth, therefore, remains the most basic necessity. The global economy has to regain momentum if it's going to be able to accommodate people and their heightened expectations with any respectable degree of dignity and human justice. Growth rates need to be higher in the developing world than in the OECD countries. This is possible for the simple reason that the developing world can utilize a lot of existing technology and apply it on an accelerated base in order to achieve that growth.

We know, however, that this doesn't need to be the traditional type of growth that we were talking about twenty years ago. It has to be more efficient, more open, more diversified. And more important, it has to be more widely shared with greater participation of the total population. In other words, along with growth must go broader distribution of economic opportunities and wealth. The "trickle-down" type of growth that was the much more accepted model in the developed world twenty years ago is not sufficient either within countries or on a worldwide basis.

With growth, the social and political problems within countries and among countries have a chance of being solved. Without it, I am not optimistic about our ability to resolve them peacefully, unless all societies enter into a period of total regimentation of the institutions that govern humankind.

International distribution of income, wealth, and economic opportunities is far easier with growth than without it. In talking about growth, we also need to be far more mindful of the environment and the quality of life than we were twenty years ago. It's not growth for the sake of machines and buildings; it's growth for the sake of humanity and its environment. But it's still growth nonetheless.

The second objective that everybody shares in the North-South negotiating process is the creation of a world of independent nation-states, the cultural and national boundaries within which we have identified ourselves over the last few centuries. The development of stable nations, in an explosive situation in which about one hundred new nation-states have come into the world in the last twenty-five or thirty years, is a crucial part of the whole negotiating process.

Developing a process of orderly economic expansion, sharing global concerns, and giving and taking in a more interdependent fashion is really what the global negotiations are all about.

A third shared objective is to achieve a world of order — or at least some kind of manageable disorder, if you want to put it that way — but certainly not an apocalyptic world.

Unfortunately, shared global objectives don't always move nations and peoples to do what would be best for everybody. So we come back to the fact that we need to identify the self-interest of individual nations and individual groups of countries in the process of those negotiations, and recognize how self-interest can be utilized to achieve the global objectives that we all have to share.

Given these shared goals, what is the present situation? It is one in which we don't have the desired kind of growth. The last seven years have been very traumatic: we have had two major recessions, the second one taking place now.

There are at least three major constraints on growth. First, growth is very limited in the OECD countries, which have the largest economies. If they grow, they help the growth of every other country. If they stagnate, they complicate the growth of every other country. Economic stagnation is now accompanied by inflation. Second, the developing world is having to face and sustain the high cost of energy. Third, there exist constraints on capital flows in sufficient quantity to take care of the economic adjustment process that is necessary for the Third World countries and their increased levels of debt.

Consequently, the implications of the current situation for the developing world could be disastrous. At best, they are quite sobering. The developing world consists of two groups of countries: the poor countries that do not have the capacity to maneuver in the face of global inflation and economic stagnation, and the medium-income countries, which many people now want to cut off from the receipt of development assistance on the assumption that they no longer need it. Yet that would be a terrible mistake, because these countries are in the middle of the economic takeoff process. Perhaps 30 to 35 percent of the population in these countries is living in significantly better conditions than before, but more than 50 percent of the population is not. For them to suffer a slowing of growth now would be fatal from a political point of view, and, furthermore, these countries may also begin to contribute to a more positive and more stable world situation if they can continue to grow rapidly for another ten or fifteen years. Therefore, any actions to cut back on development assistance to these middle-income countries must be very carefully analyzed.

These two groups of countries, then, have had to make short-run adjustments to slower growth, more debt, and weaker export markets. The question here is whether the adjustments will be too painful and too politically and socially unacceptable, or whether we can help these countries so that they can make them in a few more years in a politically and socially acceptable manner.

Many of these countries made the adjustment well the first time around (1974–76) much better than expected, and are in a position to do fairly well the second time. They need to do it in a way that doesn't slow growth much, in order to avoid potentially harmful political and social disruptions.

The main interest, then, of the Third World is to develop new forms of cooperation with the North that will complement and assist their own domestic actions. The OECD countries have a keen interest in increasing the supply of energy worldwide, since they are the greater consumers of energy. They want to have more growth with less inflation, and abundant raw materials, in addition to oil, are important to them in that respect. They are also in the best position to develop the kinds of technologies that the developing countries need in order to exploit other sources of energy besides petroleum that are potentially available in the developing countries. These are some of the mutual interests shared by the North and the South that have brought these countries together to try to develop new institutions, new policies, and new forms of cooperation within the format of the global agenda.

It is not surprising, then, that we keep returning to the same areas that have been the bases of agendas for the last thirty years, but with some new perspectives.

The area of trade is, of course, very important. And in the area of energy development, we need to finance new sources of energy.

In the area of developing new technologies for new types of energy, we need additional cooperation. Some of the new technologies will have to come from the developed countries, but others can be developed by developing countries. The Brazilians, for example, are doing a very remarkable job in utilizing alcohol from sugar cane as fuel for automobiles.

Another item for the agenda is the area of capital flows. We need longer-term and larger financial flows if the developing world is to adjust to higher energy costs without dramatically slowing the growth process, and we need to recycle part of the surplus of the oil-exporting countries back into the developing countries in order to assist that adjustment process.

It is crucial for the agenda to include not only short-term needs, but also long-term requirements. Existing institutions like the World Bank — by changing its gearing ratio, for example — can help increase that transfer of financial resources. Financing from private banks can further support the needs of the developing world.

Another crucial item of that agenda will be the search for permanent sources of financial transfers, sources that could be depoliticized. By "depoliticizing," I mean obviating the need to go to congresses in each of the developed countries every two or three years to request such financing. If we could develop such sources of financing, we could perhaps begin to institutionalize the resource flow that will be needed by the developing world for at least the next twenty years.

The monetary area needs to remain on the agenda, particularly the issues concerning monetary adjustments of individual countries and the expansion of global liquidity.

The agenda should also include food production, the technology of food production, the trade patterns of food, and the security of food availability, especially for the very poor people in the very poor countries.

One item which has not been included in the agenda here, but which has certainly been implied in the discussions of an overall agenda, is the "basic needs approach." I would prefer to call it "human resource development." As I mentioned earlier, 50 percent of the people of the developing world are younger than twenty years of age. That means we must significantly concentrate our attention on the development of those human beings — for example, in the areas of health, nutrition, and education. Had we already done so, we would be in much better shape to cope with the needs of the developing world over the next thirty or forty years.

Another item that I would suggest for the agenda is, of course, the environment. It is tragic that the United States, which for so long has been arguing for law and order in the world and has participated so well in the negotiation of a law of the seas, has, at least temporarily, taken itself out of the talks concerning the law of the seas and the environmental issues raised in this negotiation.

Another crucial need is to adapt the existing agenda to the short-run priorities of balance-of-payments adjustments, recycling, rates of growth, and energy development. Establishing priorities is important, particularly after six or seven years of frustration, which may in good part reflect the problem of an overly complex and overly ambitious agenda. It is vital to achieve some partial successes that would demonstrate we can solve these problems.

Let me, then, come to my last point: how to avoid some of the problems that have thus far prevented success in the negotiations.

One of them has to do with the appropriate negotiation process, which, as we have heard, has been a factor limiting agreement. My suggestion here is that the Cancún meeting could be a most ideal and dramatic opportunity to change the process. Perhaps at that meeting, as Mahbub ul Haq has suggested, we could simply define the areas of negotiation very carefully, but not in detail. Then perhaps the chiefs of state could delegate these issues to different committees of people who are specialists and who will be invested with power to negotiate because they deal with those issues in their own countries. For example, a committee of energy ministers should talk about energy and a committee of finance ministers should talk about finance.

A second crucial summary point that has not been mentioned but which is implicit is the ideological bent of some of the negotiations. If we could somehow first be more pragmatic in the negotiations and then be more respectful of the other people's ideology, we might get a little farther. Implicit in the process of negotiations has been a desire of one group to impose its ideological bent on the other, and vice-versa, and that impedes progress.

Another problem that we want to avoid concerns power-sharing. Power-sharing is taking place in the world and will continue to take place with growth and development, as has already happened in the last thirty years. This is a desirable thing, but in my view we don't need to negotiate right now explicit transfer of power from some countries to others. Some of that power-sharing is implicit in the results that will come out of the negotiations, and will become evident as the results of those actions begin to unfold over the next several years and as new countries manage to strengthen themselves as nation-states.

Let me now finish with several comments. The first is on the inconsistency that ideology has injected into the process.

On the one hand, countries want to be more autonomous nations. They want to enhance their options to act in many ways. On the other hand, the negotiations process has many times overemphasized the need for very specific international regulations.

If we could at least share the belief that part of the solution of the problems of the developing world will be resolved not only in trade, but in the transfer of resources, then maybe we won't need to worry so much about regulating the trade field (and with it, the problem of shifting location of industries) so much.

Finally, a regional approach to some aspects of the global development process doesn't necessarily have to mean that the South will lose bargaining power. Once an agenda is established and the issues are identified and assigned to appropriate negotiating bodies, some negotiations could take place within the Western Hemisphere, some in Africa, and others among developing countries alone. The results of these negotiations and the monitoring of progress can be done on a global basis. At least some partial de-linking could be helpful to advance the process of achieving results. All this calls for strong leadership. With it, many apparently insurmountable problems would disappear.

Chairman Sewell: I hope we will return to the issue of regional arrangements and the need for some sort of overview monitoring, because it seems to be a crucial one. Ed Hamilton will speak next.

A Political Perspective

Ed Hamilton: I have been asked to try to weave two kinds of themes together. One is the question of the United States' perspective, given the fact that the November election changed the leadership of the American government. The other is to weave the implications of that political change into the issue of overall political prospects seen from a Northern perspective.

I'm one of those who believe that the discontinuities in history in this period we are now facing may well be more important than the continuities. That is not to say that the continuities aren't interesting or that they aren't relevant to the kind of blend that is going to emerge, but I suspect that this is a period in which there will be more significant changes than there have been from decade to decade since World War II. I believe the next ten years are going to be quite interesting in this respect. "Interesting," you know, is a studiously neutral word.

Let me start by examining the United States situation for a moment, then moving from that into serious business. The United States, it seems to me, is very likely to reveal a number of characteristics that will cause the international community to view it as being significantly different in the future than it has been in the past. It is easy to exaggerate this point. I think it's even easier to minimize it, and that has essentially been the fault of a good deal of the analysis that I have seen or heard in the last six months or so.

The first truth in this area, however, is that all periods in an administration's life are not alike, and the period that is least like all other periods is the initial period. That is the period of the shake-down cruise. What is this government that we have inherited? How can we deal with the fact that the problems don't stop, but we haven't got most of our people on board yet? Who are those terrible bureaucrats who are still around, the ones we haven't been able to roust out of the corners, who are still doing the dastardly deeds that we were elected to stop? That whole set of phenomena — the business of wearing into the ruts, the new interpersonal relations, the bureaucratic personalities, et cetera — always characterizes a new administration.

The period in which all of that occurs is almost necessarily zany in many regards, as measured by either the standards or prior administrations or the period that follows this initial shakedown.

For that reason, I would not put much stock in whether the president decided to go to Cancún or, on the other hand, whether the president decided to pull the plug on the Law of the Sea negotiation in the sense that the United States backed away from its previous agreement in principle and said that it wants to look at it again.

It is not that those events don't have significance; it is just that their significance is to some degree a function of time as distinguished from the particular ideology or elaborated policy position that the administration is likely, over time, to acquire.

My own view of the Cancún decision, for example, is that at the time it was made, the president was busy trying to avoid looking like a Neanderthal, particularly vis-à-vis Mexico. If you have been the governor of California, and it is rumored that he once was, you will find that you do have a foreign policy, but it involves essentially only one country. And, therefore, you take a perspective which, among other things, does not ignore the importance of Mexico.

That situation happily coincides with objective international realities. But my own view is that the president saw nothing to lose (although some of his advisors did, incidentally) in accepting that invitation. But he has not, by doing so, necessarily suggested that a particular policy position is inevitable.

The same is true of the Law of the Sea negotiations. The problem there is that this is an extremely far-reaching treaty which was at the stage of final agreement. So the president's only opportunity to make any kind of imprint, whether he decides that he does or doesn't like it or that his people do or don't like it, necessitated pulling back for the moment and taking a detailed look at the draft treaty.

We simply need to understand that time will be required to produce anything like consistent or reliable behavior on the part of the American government.

Now let me move to my more fundamental theme. It is a sad fact, but one that must be recognized, that North-South relations seen as such — not as part of energy issues or as part of trade issues, for example — are, for the United States, *distinctly peripheral and derivative.* This view has prevailed in the United States over the last thirty years, with the exception of the one period in the late 1950s and early 1960s. At that point, because of the existence of a substantial foreign policy consensus plus a configuration of economic and political circumstances that made it advantageous to do so, some aspects of North-South relations actually became main-line politics

169

in the United States. And the president actually got elected — in part, although in small part, probably — on a platform which included a particular policy toward less developed countries as such.

Even in this one exceptional instance, enthusiasm for and the general prominence of this set of issues died very quickly. They became a victim of Viet Nam in the latter part of the 1960s. Since then, the United States has returned to the position of treating "North-South relations" as a derivative, peripheral issue. That is profoundly true now and likely to be so for some period to come, which is another way of saying that other issues are likely to dominate the administration's time and priorities. What occurs on the North-South front, as far as the United States and, to some degree, other major countries in the North are concerned, is going to be a function of other problems and issues. One might prefer to regard these things as unrelated, but in terms of the will or capacity of the countries involved to act in the North-South arena, they are going to be very directly related. Let me note two of these high priority issues, although I don't mean to suggest that there are only two.

The first issue that will dominate the stage is clearly the domestic economy and the management of Western economies in general. Members of the administration now and then intrude their personal views on the agenda, but the administration would very much like to focus on that issue. It's going to fight a very fierce battle in the United States Congress on a whole series of absolutely seminal economic policy questions. And during that period, anything that needs to be traded off or anything that is in some sense peripheral or damaging to that debate or the administration's position on that debate is likely to be sacrificed. This is really what the next six months, and probably the first two years, of the Reagan administration are about, and I think all of Washington thoroughly understands this as the main intent.

This emphasis has some very interesting implications, not by any means all negative, for North-South relations. For example, the great economic truth that has to be faced is that markets must expand if growth is going to resume. And a very important fact of domestic politics in the United States, of course, is that developing countries have become very major consumers of United States exports.

It is very likely that, absent enormous technological changes which completely revamp the market here, and absent massive changes in savings and investment rates and capital formation potential, the developing world is going to be a most volatile and important economic area for the United States. If one wants more reliable and long-term growth, one must, in fact, invest a great deal of time and attention in that set of countries.

The result is that North-South relations are seen essentially as a question of economic market expansion and the expansion of United States subsidiaries operating abroad. That perspective is likely to be much more important as years go by, more so, in fact, than it has been in the last eight years. But it is rarely related by bureaucrats to the broader, more political set of North-South issues that we're analyzing here.

The second issue that is clearly going to dominate the Reagan administration's attention to an enormous extent in both foreign and domestic policy is East-West relations. All of the things that we have talked about as causative factors in foreign policy tend to pale before the significance of changes in East-West posture.

At the moment, the United States is going through extensive internal review. I don't mean that in some organized sense a committee is thinking about the problem and will emerge on a given date with a definitive analysis, but rather that the shaking down of the administration is precisely centered around this issue: What should our general posture toward the Soviet Union be, and what are the implications of that posture with respect to the whole range of important issues, not the least of which is the domestic budget?

If, one the one hand, one is attempting to reduce federal expenditures as part of a general economic policy approach and if, on the other hand, one is attempting to increase defense expenditures, then obviously something has to give, and that something is the domestic side of the budget. The domestic side of the budget is about 93 percent uncontrollable in any given budget year. And that figure declines to only about 70 percent even if you can assume major changes in the short run. The result is, therefore, that the pressure on domestic spending is very substantial, particularly domestic spending that supports state and local activity.

This pressure results in a kind of politics, in terms of the orientation of interest groups in Washington, that is different from what we have had for some period of time. This sort of politics in turn affects the capacity of people who are interested in bettering North-South relations to develop significant constituency support on issues like foreign aid, trade preferences, and similar measures. It is totally implausible to believe that one can somehow view North-South relations independent of these new administration priorities and their broad political impact. If one makes policy judgments without having some view of what is going to happen in the priority areas, then one has no judgment. One simply has a determined sense of what is not important.

Now, given that these new priorities will condition a great deal of what goes on, what set of attitudes is likely to characterize the United States administration in the North-South arena?

First, my hunch is that there will be a lot of very tough talk. There will be a feeling that we have been carrying this development burden for a long time and that we're now going to look to our own interests because these ungrateful people in the South have been kicking us around, and we're not going to take it anymore. And there will be a number of gestures designed to demonstrate this new no-nonsense, hard-nosed attitude.

Now, to some extent, this stance reflects deeply held beliefs and is not all empty gesture. Indeed, one wishes that people were more conscious as to when they are gesturing and when they are sincerely expressing their point of view. This is not likely to be an enormously sophisticated administration with respect to balancing human values against traditional self-interest.

There is also a degree of sophistication, as several of the major foreign policy appointments suggest. Yet, even where there is sophistication on the North-South range of issues, it is heavily informed by the point of view that the best thing you can do with North-South relations is to leave them alone, that resource transfers from North to South are often based on intellectually bankrupt foundations, and that, in this arena, "less" is far better than "more." Whether one likes it or not, this point of view is very prevalent among people in the new administration who know something about these issues and will have significant influence on policy.

Incidentally, presidents always inherit the less developed world as a relative surprise. When they wake up on January 21, there are suddenly 120 relatively unknown countries to think about. Immediately they're overwhelmed by cables from all over the world and ambassadors asking difficult questions and state dinners for a confusing array of foreign dignitaries whose names they can't even pronounce. A president suddenly has to focus on a whole area that he hadn't considered before. So some of this administration's hardball attitude will be laced with some charming naivete.

Nevertheless, after the initial neoconservative reaction, I think a second characteristic will be a major attempt to differentiate among less developed countries. An effort to take on the Group of 77 as a whole and to follow the route of "global" negotiations is not likely.

The one exception would be if this negotiation somehow becomes part of a package which is perceived — even if it's not stated as such — as part of a general solidarity movement having to do primarily with East-West relations. If negotiations become a function of that issue and if the United States then found itself pulling people together on all fronts in an attempt to maintain a united stance, then one could predict a quite different situation. But the initial instinct of the people who have just come to office is to differentiate between groups of countries and between sectors, and not to try to carry on general negotiations.

A third characteristic will be a heavy element of mercantilism in United States policy. That is, the United States will concentrate on narrowly defined self-interest, weighing short-term economic gains and losses very carefully. This emphasis on mercantilism could mean several things. It could mean a reinvigorated GATT, essentially along classic lines, or it could mean the steady, continuing erosion of the GATT if the United States decides that the best way to tilt trading rules in its favor is to establish them in novel kinds of bilateral and multilateral configurations around the world.

It is very likely that there will be considerable internal tension in the American bureaucracy between those representing commercial interests and those representing foreign relations in the new administration. That will not, of course, distinguish it from some earlier ones. Nevertheless, it's important to understand that the American government is going to have a difficult time speaking with one voice on a wide range of foreign economic policy issues, a rather more difficult time than under recent U.S. administrations.

It is also worth remembering that Ronald Reagan is the one man who can get away with a substantially pro-LDC policy position in current American politics should he decide to do so. Given his position at the right of the American political and ideological spectrum, he has more freedom of movement — as did Nixon vis-à-vis China — than any president coming from the left of center.

Now I'm going to turn to a more general set of points.

First, on the question of global institutional structure and the interesting dilemma raised by the comments of Mahbub ul Haq. Mahbub is an extremely shrewd and acute observer and predictor, and he thinks that three institutions, designed to perform functions that were given major consideration at the Bretton Woods conference, will be developed in this decade. That is to say, we will have a world development fund for long-term development financing; we will have a world central bank, which is what the IMF was supposed to be; and we will have an International Trade Organization which has the characteristics that the delegates at Bretton Woods could not agree on and, therefore, which was not created.

This idea represents an enormous step in global institutional architecture. Yet it is suggested at a time when such a step seems out of the question. Indeed, trends in international economic relations — including relations between developed and developing countries — appear to be moving in the opposite direction, away from the present degrees of "globalism" inherent in our international institutions. We clearly need a more informed notion of where we are headed in the next ten years.

My own hunch is that Mahbub is substantially more optimistic than the events will warrant, barring some very cataclysmic events that might indeed produce the kind of institutional renovation he's talking about.

On the other hand, one might move toward one of those innovations. In the area of trade, for example, the present juxtaposition of disintegrating rules and norms on the one hand and great and growing degrees of economic interdependence on the other could well lead to some institutional innovation in the direction Mahbub predicted, though stopping well short of his ITO goal.

The same can be said, with perhaps more confidence, on the international monetary front because in this area, major crises can develop very fast and produce a kind of "shivering of timbers" among national policymakers that is very hard to produce in the trade area. I doubt strongly that any major institutional innovation will occur in the area of a world development fund.

A final point about organizational architecture. The interesting organizational issues still revolve around the question of how one manages the process of international economic change, not whether one is able to produce an institution designed to take on the issues as though they could be resolved once and for all.

The issue of legitimacy of international institutions and rules has not been discussed as much as it might have. As someone has said, the minute the issue of legitimacy enters the negotiations, everything becomes nonnegotiable. I would argue, however, that legitimacy arguments are an inverse function of urgency. That is, the more urgency one feels and the more one's fundamental survival or comfort is threatened, the less we experience the kinds of sensitivities that prevent two parties from even talking to each other.

I suspect that on a series of sectoral fronts, such as energy and food, the urgency will be so great that the question of legitimacy, which cannot be directly negotiated, will be implicitly negotiated. There will be a jockeying that will result in a sovereignty configuration different form what the negotiation began with, but probably at the price of never mentioning it. This is going to be a difficult thing to do, particularly for the South. The North is arrogant enough that it doesn't feel the need to talk about its sovereignty so much.

A final set of points I'd like to make concerns the likely *substance* of the North-South agenda. I tend to believe that the negotiations will, in fact, be sectoral, at least in the short run. That is to say, any attempt to produce a broad compact, whether it's called the new or the old international economic order, will be essentially derivative. The main activity, the engine of change, will be a series of sectoral negotiations. Those sectoral negotiations will be more regional, or at least more bilateral or selective among countries. Negotiations between the developed countries and the NICs serve as an example of what I suspect will develop. The approach of sectoral/regional/country grouping will become more important, regardless of whether the administration decides to announce that it will occur.

The issue that is most likely to confront us in the short run is debt. I suspect that we will have the tremors of some fairly serious debt service questions among very large debt service carriers — Brazil is a good example, but there are others. This will, again, "concentrate the mind" in a way that other issues in the international financial community don't effectively do. And debt is probably the vehicle by which more less developed countries can be brought into some kind of negotiation, either directly or indirectly, than virtually any other issue. It's also the one that, in terms of Northern politics, it is easiest to argue, "can't be avoided." You can always argue in the United States Congress that a foreign aid bill or some sort of trade preference action or a customs bill having to do with a particular product or quota can be put off. It's harder with debt, mostly because bankers are such unforgiving, deadline-oriented people. They have the quaint view that time is money, and they tend to make that issue very clear when this subject arises in Congress. So I would expect debt to be a substantially more important issue than it has been in the 1970s.

On the issue of direct foreign assistance transfers, or commercial transfers, my hunch is that United States policy and Northern policy in general — not including the Scandinavians, the Dutch, and perhaps the Canadians — will be distinctly derivative of export promotion. Economic policy will be seen from the point of view of narrow national advantage: export promotion on the one hand and capital remittances on the other.

Therefore, the direct transfer issue, which is usually embodied in the foreign aid bill, will become much more quiescent than it has been. Of course, there will be incendiary activities in Congress; there always have been. But they will be about small amounts of money compared to the total flows between developed and less developed countries. Of increasing importance will be a series of relatively invisible measures concerning particular trade negotiations with particular countries.

Again, I would urge you all to keep your eye on the East-West situation. This is going to condition a good deal of what is found to be possible in Congress and in the body politic during this period.

To summarize, the discontinuities that I have described are on balance going to outweigh the points of continuity, at least for the next couple of years. We will have a shakedown period for the next six months or so in which it will be hard to tell what's continuous and what's discontinuous. After that, habits will settle in and, as a result, behavior will become somewhat more predictable.

Chairman Sewell: Ed, thank you.

We're fortunate in having with us Jorge Navarrete, who will have a large voice in the preparations for the North-South summit scheduled for October in Cancún. Jorge, I wonder if you would like to say a few words about the real world.

Jorge Navarrete: The proposed meeting in Cancún, Mexico, in October of 1981 is called the International Meeting on Cooperation and Development, and is to be held at the level of heads of state. It is the common initiative of a group of eleven countries, six developing countries and five developed ones, whose foreign ministers have so far met twice in Vienna to define the main objectives of the meeting. So far, this group has defined some of the major features of the meeting. Others it has left to be developed later.

It is agreed that the meeting will be an informal one, encouraging a free interchange of points of view among the participating leaders. In no way is it being conceptualized as an opportunity for negotiations. Indeed, there is no mandate for participants at this meeting to negotiate in the name of anyone. The participants will not claim to represent groups or regions or any other collectivity in the international community.

The total number of participants will be restricted in order to accommodate and encourage free and open interchange. It is agreed that there will be twenty-two or twenty-three participants.

The meeting will not try in any way to preempt or substitute the formal process of multilateral economic negotiations in the United Nations or in other forums, but will attempt to contribute to these formal efforts of negotiation, particularly in relation to the Global Round of negotiations.

If there is some kind of meeting of minds, some degree of consensus at the meeting, this would be a positive step toward continued and more fruitful efforts of formal negotiation.

It is widely agreed that these discussions in Mexico will take the same general approach that is behind the concept of the global negotiations in the UN. This means that no issue will be isolated or singled out from the others; the discussions will try to see the problems of the international economy in their complexity and take into account the interrelationships among the issues.

Being an informal meeting, there will be no formal agenda for the discussions, but some sort of framework for discussion will be proposed to the participants. This framework for discussion will encompass the main issues on which the interchanges might be concentrated. Perhaps some optional approaches to policy as guidelines for future discussions and for future action will be offered to the participating leaders.

So far the eleven countries have not defined the main items to be put within this framework for discussion, because they have felt that this is a task for all of the participants and not just half of them. But there is an initial, informal understanding that this framework will include four or five main areas of concern. We still need to discuss how these areas of concern are to be defined and what issues will be highlighted within each of them.

The four or five areas of concern that have so far been selected are the following. The first is food and agriculture. In any North-South discussion this problem has to be addressed: not only the short-term issues of food security — the problems of starvation in different areas of the Third World — but also the longer-term issue of agricultural development in the developing countries. If just the short-term or just the long-term aspect of the problem is taken, the approach will be incomplete and important elements that must be addressed will be omitted.

In the second area — trade and industrialization issues — this same approach is also adopted. It addresses the present short-term problems of trade, particularly the question of the reemergence of protectionist tendencies. The longer-term perspective focuses on the industrial development of developing countries and on the reindustrialization or restructuring of industrial capacities in the developed countries. Perhaps this approach will provide the opportunity for an examination of questions of mutual interest that concern immediate as well as longer-term objectives.

A third area of concern will be energy. Here, I think, the approach is still less defined than in other areas. The problems of energy-deficient developing countries will have to be given a high priority, but there is yet no clear perception of how to connect these with other energy issues.

The fourth area will concern financial and monetary questions, and here also the two-fold approach seems to be very clear. In the short run, there is a need to respond to the necessities of the adjustment process, to the increasing deficits of a great number of developing countries. And in the longer run, there is the need to respond to the increasing need for the transfer of real resources to developing countries and to tackle the problem of the long-term debt burden of many developing countries. It is also important to discuss some crucially related institutional issues: what kind of evolution or reform is needed in international economic institutions, particularly in the monetary and financial field, but also in trade and other fields.

Some people involved have suggested a fifth area of concern: interrelationships between the use of natural resources, population growth, and the environment. This is also an interesting area of concern and one that will connect the discussion with some of the most pressing social problems in the world community.

The process of preparation is not yet completely defined. The foreign ministers of the eleven cosponsoring countries and of the other participant countries will meet once more before the October meeting, in early August. From now until August, all interested countries will continue the process of preparation, focusing on these four or five major areas of concern. In August, when the topics are more clearly defined and the framework for discussion agreed upon, perhaps a more systematic process of preparation will start.

It's difficult to define briefly the kind of results we are aiming for. Everybody generally agrees that the outlook for this kind of exercise is not particularly bright at the moment. In a situation of slow growth, inflation, and unemployment in the developed countries, the incentives for improving international cooperation for development are not the strongest.

But given what this situation means for the world economy, the possibility of resuming growth in a more integrated and interdependent world economy is one reason why this may be the right moment to discuss and try to improve the measures of cooperation.

The discussion will, I think, be political in character, because there will be political leaders around the table. A possible result of the meeting is some degree of consensus that could be useful in reviving and guiding formal negotiations now stalled in other forums. The meeting could improve the present environment of international cooperation. It could result in a better understanding of different perceptions of the problems on which there are still important differences in approach or in procedure.

Hopefully, the preparation process will offer the meeting participants options to discuss. We are not attempting in any way to draft a declaration for the leaders to approve and present to the world. Rather, we are trying to give them several alternative approaches to different problems and to see whether, through the discussion, one of these approaches could be endorsed, or whether a combination of these approaches is the most adequate solution, or even whether alternative options could develop from the discussion itself.

All the participants, and particularly the host country, will make a definite effort to prevent the rise of overblown expectations about the possible results. The problems are difficult. The situation is difficult. Simply creating the opportunity for responsible political leaders to meet and to discuss these vital issues freely is worthwhile in itself.

I would like to conclude by returning briefly to an issue which I began to develop briefly yesterday morning. Yesterday we more or less agreed that, in the UN and maybe in other international forums, the process of negotiation itself is sometimes an obstacle to understanding and to results.

I agree with this observation, but for different reasons than the ones so forcefully presented by Joan Spero. She mentioned that in New York the Group of 77 speaks with only one voice, whereas the Western developed countries act individually and quite often send different messages to the developing countries. She even said that sometimes this is a cause for confusion on the side of developing countries, which are not clear as to what course of action is favored by the developed countries.

But maybe this is not just a circumstance of the negotiations. Maybe it's a tactic of negotiation on the part of the advanced countries, which find it easier to send different signals and to respond from different angles to the unified voice of the developing countries. As a result, on the side of the developed countries the point that is finally taken is the lowest common denominator. So it's an exercise in which the most positive positions of developed countries do not normally prevail.

A further problem with the process of negotiation seems more or less peculiar to the United States. It relates to the issue of coordination among the different government agencies that participate in the negotiations as part of the United States delegation or that present opinions on issues to be negotiated. Here again, the least common denominator principle seems to control and explain the final positions taken by the U.S. delegation.

I have two other minor points. First, it is clear that the diversity of interests and of levels of development within the Group of 77 gives us an initial position that is sometimes not very well defined or completely coherent. But it is also true that, during the process of negotiation and in the interchange with other countries and groups of countries, this initial position is clarified and made much more coherent through the consultation mechanisms of the Group of 77.

Finally, Joan mentioned that this interchange between politicians and diplomats from the Third World and economists from the developed countries is not always a fruitful dialogue, and that sometimes it's difficult for one group to understand the other. I think that maybe the real problem is that sometimes the analytical tools of the economists representing the developed countries in these discussions are not the most adequate to apply to the developing societies whose problems are being discussed or negotiated.

Panel Discussion

Chairman Sewell: It seems to me that at least three separate issues have been raised. I would like to hear brief responses to them in the framework of this afternoon's topic of discussion — that is, the possible shape of the agenda.

The first of the three issues is about process.

There seems to be some concern about the efficacy of the Global Negotiation at the United Nations and a desire to see it move into functional and/or regional subgroupings, where the whole process would be much more manageable. Of course, we have seen this happen several times, most notably in CIEC, without much result.

It will be interesting to get people's reactions to the process argument, particularly to the question of how you combine disaggregation with a global overview. Gerry Helleiner very eloquently talked about the need for a global "beacon," because most subglobal negotiations tend toward the short run on a particular functional problem and not toward the global interest. We need to consider the trade-offs on that kind of issue.

The second is the issue of substance. We tend to revert to what I will unfairly call the traditional set of negotiating issues, and they are seemingly reflected in the planning for the Cancún summit. Are there other ways to cut the issues and, if so, what are the trade-offs?

If we adopt the UNIDO goal — which raises hackles all over the North — of industrializing the South so that it represents 25 percent of the world's industrial production by the year 2000, what is the trade-off for reindustrialization of the North?

Is it really more efficacious, for instance, to talk about recycling OPEC funds into the South, which seems to be ultimately desirable and practically impossible, than to recycle them into the North? If the United States automobile industry needs $80 billion of investment, maybe it's better to put these funds into that industry and thereby acquire the freedom to do a variety of other things in resource transfer, trade liberalization, and so on.

Is it really time to talk about some agreed-upon level of protection in the North? It will happen anyway, and it may even be desirable, but the trade-off would be much greater support for foreign financing and cooperation in the South.

In other words, are there ways of shifting the negotiations out of their usual mode and basing them on Northern self-interest? Unless negotiations are based on Northern self-interest, whether it's in the narrow East-West framework or in our own economic context, I don't see the negotiations going anywhere

Finally, I would like to return to the very interesting point that Professor Rostow raised at some point yesterday: an informal consensus among experts from the North and South of the kind that existed in the late 1950s and 1960s and that perhaps gave rise, at least partially, to the political climate that Ed Hamilton referred to. Why don't we have such a consensus? There has been a rash of diverse reports emanating from both United States and international groups, such as the World Development Report, the reports of the International Monetary Fund and the GATT, the Brandt Commission Report, and our own Global 2000 Report, upon which such a consensus might develop. Why hasn't one appeared?

On the other hand, is a consensus really a precondition to action? If all of us can so eloquently describe the mutual benefits that can be achieved, why doesn't anybody else agree with us, at least in a political sense?

I would like to see comments on these issues, because it seems to me that they are some of the problems that underlie the actual agenda for negotiations at the global level or any other level.

Walt Rostow: I want to share a bit of my experience from the summit at Geneva in 1955. There sober staff people and foreign ministers met and developed an extraordinarily clear-cut, intelligent proposal for all the chiefs of government: namely, that they discuss issues but refer substantive matters to their foreign ministers for negotiation. It was Secretary of State Foster Dulles who insisted on this, and it resulted in keeping everybody but Foster Dulles out of Geneva.

My guess is that when real live working politicians sit even for two or three days on the world stage, they are going to try to say something of substance. The dynamics will force them.

President Eisenhower put forward, to Mr. Dulles's initial discomfiture, the open-skies proposal for mutual aerial inspection. It was not a gimmick. He deeply believed, to use his own phrase, that this was a small gate into disarmament. Mutual aerial inspection has come to life, and it's one of the few stabilizing instruments that we have left in the arms race.

In any case, don't be surprised if our politicians get a little romantic and ambitious when they actually arrive in Cancún.

Second, responding to what Mr. Navarrete and Mr. Barletta have said, I would like to make a contribution to Mr. Navarrete's useful observations on what the summit might accomplish. I was much moved when Ambassador Mills said yesterday, in response to the raising of the functional/regional approach, that if we give up on the global approach now, we may lose the New Economic Order forever.

For reasons that I have expressed, I am not sure that it was a good approach to begin with. On the other hand, it reflects some deep feelings and commitments. Therefore, the central task of the summit is to find a way to sustain the continuity of the New Economic Order debate, which Mr. Barletta referred to as "educational" and which reflects strong political and human forces. Let it go on. It should not die. And if people believe that it's going to have fruitful results, it will go on and it should go on. But the New Economic Order must not make itself the enemy of getting on with the tasks of growth and of dealing with such mortally dangerous problems as food, energy, debt rollovers, and other urgent issues. They may sound pedestrian, but they are certainly not pedestrian to the human beings involved and for the developing countries whose growth has been slowed down by our failure to deal with them.

What would you assign to a global level? You would assign to a global level whatever people want to argue about in the New Economic Order. But there would also be certain functional tasks. For example, trade obviously has to be dealt with at a global level. The same goes for commodity agreements, to the extent that they are rational, and new gearing ratios in the financial institutions. Rollovers are essentially global because they involve Japan, Western Europe, and the United States. Those are functional issues, beyond which you would provide for however much continuing UN debate on the New Economic Order as the South desires.

But the resource issues and the problems of small vulnerable countries are, I think, better administered regionally for four reasons.

First, as far as the sectors are concerned, we are talking about major investment programs. You know your resources better in the region; and it's a more manageable group of countries to deal with.

Second, there are regional development banks, which are valuable institutions. A good deal of this resource development activity could be centered around them and backed by the World Bank.

The third reason is that in your own region you have a better knowledge of the poor countries and the help they need. You'll get more help from Europe, let's say, if it is leading the way with Africa. You'll get more help from the United States if it is leading the way for poor countries of this hemisphere than if aid to the poorer countries is all part of some big global Waring blender.

Finally, a regional approach leaves greater scope for South-South cooperation, which is going to become increasingly important. It's really one of the most fundamental ways for the South to gain greater control over its destiny.

Now, let's look at some of the functional aspects of global arrangements. For example, certain aspects of raw materials might be dealt with globally. Biomass is not something invented by Brazil or relevant only to Brazil in the Western Hemisphere. There is no reason why you shouldn't provide for global discussions of the use of alcohol and other forms of biomass.

A continuing serious forum could also provide for a global monitoring of the whole set of activities. I'm not talking now about some central committee that gives orders to the World Bank or to the IMF or to the Inter-American Development Bank. They each have their charters.

So you would have at the global level a continuing New Economic

Order dialogue, and you would have a serious monitoring of all these activities. But then you would let the hundred flowers bloom. You've got to. These are such big issues. If you try to do them all at the global level, nothing is going to get done.

Now, finally a word in response to Ed Hamilton's rather brilliant exposition of the new administration in its early days. I'm inclined to agree that it's important to understand its initial stance. It's quite reminiscent of the early days of the Eisenhower administration. Only after considerable trouble did the Eisenhower administration take the South seriously, and it did so only in its second term. But the present administration began with a major crisis in the Caribbean and Central America. As you know, you need a lever to move the world.

It's obvious that this administration is going to need a positive program for that part of the world. I don't know what precise problems are related to stopping the external flow of arms that is going on. But obviously we cannot in reality just devote ourselves to countering terrorism while a lot of those countries are feeling the effects of the functional problems that we are talking about: high oil prices, excessive food imports, economic slowdown in the Untied States, et cetera. So we're going to need a positive aspect to that program fairly soon. This may be the beginning of the road that leads the Reagan administration into a constructive North-South policy.

Chairman Sewell: Gerry Helleiner.

Gerald Helleiner: My question is whether the summit is, as somebody put it, something with a beginning and an end, or whether it can be viewed in terms of process. The Western economic summits are part of a process by now. They are held on a regular basis, and, in fact, the North-South issue is high on the agenda this year. As you know, there will be another one, and another one, and another one.

Is the Cancún summit seen by the organizers as the first of a series, a new process which supports or serves as a backdrop for the Global Negotiation, or is this a unique event?

Jorge Navarrete: I think that everybody representing the eleven cosponsoring countries recognizes that this is part of a process. Certainly other meetings similar in organization or in character could be held. But, of course, this is not in any way the beginning or the end of North-South discussions, we hope.

Gerald Helleiner: But the question I'm really asking is about institutionalized, perhaps annual, North-South summits.

Jorge Navarrete: I think that question is a difficult one to answer. Annual Western summits are now a feature of Western economic relations, but I'm not sure whether this other kind of summit meeting could also be a more or less regular feature. I suppose it very much depends on the results of this one, which is not the first. Several years ago in Jamaica, there was also a similar exercise.

Chairman Sewell: Gerry, did you have other comments?

Gerald Helleiner: My other question is addressed to Nick and to Ed.

I was made a little nervous by Professor Rostow's remarks and by remarks I have been hearing in other places about the possible regional character of the North-South hardball that the Reagan administration may introduce.

It's extremely easy to slide into the old sphere-of-influence view of the world. The Europeans' clear preference is already represented by the Lomé convention, which is not such a great achievement as you were suggesting, Mr. Chairman. And they have now created a European currency unit which is clearly intended to serve as the basis for their particular sphere of influence as a challenge to the dollar. The Japanese have a history of similar approaches.

Now, I must confess that I'm a little nervous to hear the World Bank's vice president for Latin America (and last week the chief economist in the Inter-American Development Bank) making "Western Hemisphere" noises for the first time in twenty-five years, and quite contrary to the warnings in the early 1960s of Raul Prebisch.

There may be sympathy on the part of Latin American countries for a set of Western Hemisphere arrangements because nothing else is possible, and I'm not sure that that is the best way to think about organizing the world.

Walt Rostow: That's a pretty good reason to think of something else.

Chairman Sewell: It seems to me a central question. I'm glad it has been clearly posed. I would like to hear other people's views on it, too.

Paul Streeten: I agree with Gerry, but he should spell out his reasons, which are not so immediately obvious for everybody.

Gerald Helleiner: There are self-evident reasons in terms of the sorts of long-term developing country objectives Don Mills has been putting before us. A variety of Southern countries with more or less common grievances in the whole North-South discussion are risking their cohesion in attempting to make unilateral and/or regional gains. They may not always do that, but it certainly can be a potential problem if each smaller group deals with one major power. It's a version of the case-by-case approach. It may come out all right, but it has the risks of Northern "divide and rule" strategies, the carving up of spheres of influence and the cartelization of markets.

The French still receive higher prices for machinery in their former colonies than anybody else. These things are insidious, once started, and this may be a perfect example of the possible negative consequences of pursuing what appear to be short-run beneficial policies in the belief that this approach works best in the long run. It might be that no action is better than this kind of action.

Chairman Sewell: I would like to stick with this issue of regionalism for a bit and hear from Don Mills.

Donald Mills: Some time ago the same issue came up in an informal discussion about the North-South negotiations. In the middle of it, one official said that recently Jamaica and his country had a very good bilateral meeting, and that lots of good things could flow from it. He went on to suggest that it's better to deal with our issues using a country-by-country, case-by-case approach.

My reply was that there is every good reason to have good bilateral relations, but that's not what we're talking about in the North-South negotiations; you cannot change the system in that way. You don't change the cultural values which have the greatest significance, particularly in light of the history of the last three or four hundred years.

And I really feel strongly about this. It's a history in which one sector of the world — which is becoming a shrinking sector in terms of population, not in terms of economic or military power — has dominated the rest of the world for a very long period.

For the world to reach a different level of accommodation among its peoples, we have to change some of these values and the economic and other relationships associated with them. You don't change them by establishing paternal relationships, whether on a regional basis or otherwise. So fitting bilateral relations into this attempt to establish a world community based on a different set of values in a different set of relationships is the challenge that we face.

So let us have our bilateral relationships, and even things like the Lomé convention, which has some value but which should not be overrated. But let us not have what has been pushed at us so frequently by the Trilateral Commission and others — the idea that economic relations in the world should be pursued on a basis of regionally organized interests centering around the industrial countries. To achieve fundamental changes in global structures and relationships, we must put the interests together and the issues together.

So the challenge is to accommodate bilateral relations but also to recognize that in certain areas a global perception must be the basis of the establishment of a different type of global community.

We must approach some of our problems globally while at the same time recognizing the need for action in other spheres. There are some things that are far better done outside the global forums and there are some things that cannot be done in the global forums at all. We have not given sufficient attention to this distinction, or to the fact that some global efforts flounder precisely because of lack of attention or commitment at the regional or national levels.

We have mentioned the private sector — bankers and commercial interests, trade unions, and others. But we don't talk to them in the North-South discussions, and they don't talk to us. We have not created the institutional framework whereby these interests, which are vitally concerned in the same issues, can come together. So there are gaps in the system which we must mend.

Chairman Sewell: I want to stick with this issue of what requires less than the global forum.

Paul Streeten: My major concern is that the regional approach neglects India and Bangladesh, where some of the poorest people live. Let me elaborate a little.

The regional approach clearly produces alignments like that between Europe and the countries of Africa, the Pacific, and the Caribbean. They have already got together and have made arrangements that are in some ways beneficial to Europe and in some ways beneficial to them. Japan is already creating a coprosperity area with some countries in East Asia, and would presumably look after its interests. Perhaps they would benefit from this. The United States would make some kind of an agreement with Latin America, Central

America, and possibly the Caribbean.

But if we're concerned with poverty, we ought to be concerned with the Indian Subcontinent. And I cannot see any country particularly interested in making agreements with the Indian Subcontinent. It would be left out, and the concern with poverty and basic needs would go by the board. This criticism applies not only to the regional and subregional approach, but also — and let me say it once again — to the sectoral and functional approach, attractive and interesting though it is for other purposes. And it also applies to any narrowly interpreted mutual interest approach.

Take the energy issue as an example of the sectoral approach. If we get together and arrange for investment in new oil exploration and oil substitutes in some countries, we may discover oil deposits in some very poor country, perhaps in Chad or Mauritania. Lots of investment might go on there, but the impact on genuine development in the world would be utterly capricious and entirely coincidental. It is important what interests we mobilize politically for these actions. They would, again, be rather partial and functional, and the main issues of poverty and basic needs, which call for a multisectoral approach and for a different mobalization, would be neglected.

Now, I suppose several things could be said in reply to this position. In the last twenty or twenty-five years, we have perhaps unlearned more than we have learned about development. As Mark Twain once said, "The trouble is not what people don't know, but what they know that ain't so." But if we have learned anything, it is surely that development is a complicated multidimensional affair in which a broad range of actions of the developing countries and the international community have their impact on the development process. All of them have to be taken in order to achieve certain results. In fact, action on only one front can either be futile or actually counterproductive, enriching some small subgroup in that country and leaving whole areas of that country untouched or impoverished.

So if we are serious about poverty eradication, development, or growth with equity, we ought to look at the impact of all these actions, not only in the areas of energy or food or raw materials. And unless we do this, *we're really doing something for quite different purposes than the support of development.*

As for procedures, of course, it doesn't take much experience or intelligence to know that you can't talk about everything at once with everybody; the world and the issues are far too large and numerous. Even the Congress of Vienna was quite difficult, and the world was much smaller then. And I'm fully aware that we must negotiate about specific issues, with particular groups, in a time sequence.

However, this type of negotiation does not preclude the possibility of bringing the partial results of regional, sectoral, and functional negotiations together under one framework. This would allow us to understand what we have left out, what we have neglected, and what the impact is of these different pieces we have dealt with, and then to coordinate them in a more rational way. But unless we combine the functional, the sectoral, and the regional approach with the global view, we are buying progress at too great a cost.

Chairman Sewell: Paul, thank you. Ed Hamilton.

Ed Hamilton: My original charge was to talk about probabilities, and I would extend what Walt Rostow, Don Mills, and Paul Streeten have said.

The probability is that we will not sort out the global versus the regional/sectoral approach as neatly as we would like to, or in a way that would seem, as an intellectual proposition, the most obviously needed. But there is a lot to be said for the proposition that in the next ten years the primary task of statecraft is to develop an ongoing process that concerns itself with the global issues and that avoids the partial approaches that are really harmful, the kind that Gerry Helleiner and Paul Streeten rightfully worry about. As I have said, moving in this direction will be an enormously difficult political feat.

We need a process that marries the global concerns with particularization wherever particularization makes sense and has the added advantage of galvanizing the will of people who aren't interested in global architecture.

Things work in cycles, and North-South relations are no different. We have been through a period in which the fashionable mode of conversation and thought has been global. We are probably about to go through a period in which the fashionable mode will be particularization. That is not necessarily the end of the world, but it certainly places an enormous stress on the institutional foundation that we have been able to establish in the past three decades. It will, therefore, create enormous problems as well as new opportunities.

Part of that opportunity is the recognition that a good deal of the New International Economic Order can't be effectively dealt with on a global basis. Yet there is a whole series of things which must be maintained on a global basis if one is to prevent an irreversible "beggar-thy-neighbor, tit-for-tat" kind of world, which will make net losers of us all. The trick, therefore, is to balance the global approach with the more particularized approach while at the same time balancing North-South relations with a number of other power relationships that are in considerable flux. This is a neat trick and one that we will probably not accomplish in a particularly tidy way. The test will be whether we can manage that challenge effectively.

Chairman Sewell: Roger Hansen.

Roger Hansen: There are two comments I can't resist making, and both reflect highly personal views and values.

As a person who started out in the field of history, I am often scared to death by the malign effects of what may well begin as benign efforts. The history of regionalism, when we think of it in terms of international relations, has indeed been the history of interests generally defined by hegemonic powers. This is true in an overwhelming number of historical instances, and it seems that there is a degree of historical inevitability to this outcome. So I instinctively shy away from a regional approach as a value preference; I have the feeling that weaker partners will tend to get the short end of the regional stick.

But beyond that, in an objective sense, I suspect that present degrees of interdependence — *dependencia* theorists would still call much of it "dependence" — are such that a regional approach to resolve many issues will very likely fail even if it's tried.

If we're thinking of nuclear nonproliferation, what does "regionalism" mean? If Brazil wants to buy full nuclear fuel cycles and West Germany will sell them, what does the approach of regionalism contribute?

If we're worried about the global food problem and look at the vital global food trade patterns, I don't see much that comes out of regionalism. Rather, I see a desperate need for the evolution of rules and regulations and incentives that are almost wholly global in nature.

Again I fall into the Paul Streeten camp in saying that in one sense we can use whatever progress we can make on all fronts. On the other hand, I suspect that both the nature of the issues and my concern that Northern partners will ultimately define the content of regional programs involving both developed and developing countries make me somewhat cautious about regionalism. My suspicion about Northern dominance in regionalism is certainly shared in much of the South. It is clearly known that the United States is wanted by the ASEAN countries in any Pacific Basin arrangement as a counterweight to Japan. So you have the United States in the Pacific; you have the United States in Latin America; and you obviously have us in the Middle East and Africa as well. As for South Asia, let me simply underscore the Streeten concern with the potential human cost of regionalism in overlooking needs in that area.

It seems to me that the games of interdependence have gotten away from us at this point. We are so globalized in so many ways — economic, political, cultural, and social — that regionalism simply may not be able to carry us as far as it might have twenty-five years ago. More important, perhaps, it may well carry us in the wrong direction.

Chairman Sewell: Nick Barletta.

Nicolas Barletta: I think it is extraordinarily interesting to see the immediate gut reaction of some people when one mentions some of these things. I can assure you that Latin Americans are not thinking of spheres of influence with the United States. But it is very interesting that this attitude was the first one to crop up.

In Latin America, the Brazils, the Venezuelas, the Mexicos, and the Argentinas are going to play an increasingly significant role in the world arena whether the United States likes it or not. They have the capabilities, the size, and the potential power, and they are going to do so. The Andean group, if it continues to develop, may do the same thing. So we don't have to accept the idea that the United States is trying to impose its will on Latin America.

The key point here is to recognize that bilateral exchanges represent more than 95 percent of what's going on in the world whether we like it or not; and global approaches really don't add that much to the solution of the world's problems right now. That may be unfortunate and tragic, but it's true, and we shouldn't avoid this truth.

I am willing to submit that the greatest impediment to constructive action in recent years has been the global approach to negotiations. Toward the end of the 1960s and the beginning of the 1970s, the Latin American countries had the United States with its back against the wall in hemispheric negotiations. I do not think that was necessarily a good approach either, but that was the case. And the United States escaped from Latin American pressures by moving to the global arena. And in the world arena, what has been happening?

Nothing. The discussions became too broad and complex to achieve concrete results. So let's turn all of these things around, because it's interesting to look at them from different angles.

From one point of view, and I think it's a valid one, the Third World can exert some degree of power when it gets itself together on some issue. But it tends to be the lowest common denominator for all. And even though everybody may agree with that position, out of decency and humanity and a genuine belief in cooperation, the Mexicos and the Brazils and the Venezuelas and even the Nigerias and the Indias are going to take care of their own problems first because that is their first duty. The world is organized around nation-state sovereignty, and countries will take care of their problems anyway and in cooperation if possible.

Another point that I would like to mention concerns process. It has been unwieldy to take on so many issues of such vast importance on a worldwide basis with so little power to deal with them. If people who have been trying to solve these problems in these forums have not had the power to deal with them, there has been nothing to do but just keep on talking.

What we are proposing is not globalism versus regionalism. Let's combine the different approaches in a way that will produce results, but not in a way that will delay solutions for ten more years. The problems here are festering, and they must be solved in some way.

The combination that Walt Rostow proposed, I think, is an interesting one: a global monitoring of certain things that need to be handled on a global basis. The United Nations may be the best forum for that, with an annual meeting, Cancún-style, of a small group of chiefs of state. But we can delegate sector problems to specialists who have the power and the knowledge to negotiate and deal concretely with the issues, after considering the benefits and costs to each of the participants. They can perhaps produce concrete and clear solutions, because they are specialists in that subject.

Walt Rostow cites a very significant problem, the plight of the little countries in the Western Hemisphere. Fifteen small countries in the Caribbean and Central America are having tremendous problems. And who is going to be concerned about those problems? The Africans and the Asians and the Europeans aren't concerned; those most concerned will be the people living right next to them, the Mexicans, the Americans, the Canadians, the Colombians, the Brazilians, and the Venezuelans. It's only natural that it should be so. Why should that problem, then, go into the global arena?

Roger Hansen: But let's return to Paul Streeten's question, then. Under this approach, who helps the three-quarters of the world's population on the Indian Subcontinent?

Nicolas Barletta: That remains a global issue. The whole issue of the poorest nations may need to remain a global issue, and not necessarily in the United Nations, if we are to solve some of these key problems.

If we manage to change the gearing ration in the World Bank and if we manage to find other ways of channeling resources through existing institutions, we can continue to provide a lot of effective financial assistance to the very poor, which is absolutely necessary.

Let me just mention two or three other points.

In the whole issue of South-South negotiations, why do we have to gear everything to the OECD countries? There are certain things we have to do in cooperation with the OECD countries, and there are certain things we have to do among ourselves. The biggest problem of dependency, for those who like to talk about dependency, is the mental one. There is no reason whatsoever for depending so much on the OECD if we are mentally able and willing to act on our own. When we used to telephone from Brazil to Caracas, we had to go through an exchange in New York. Yet all we really had to do was install a direct telephone connection between Venezuela and Brazil.

Many times we are not aware of our powers and our capacities to do things for ourselves. We keep hoping that the OECD will solve our problems for us. I think this is part of the implicit ideological problem which is limiting our success.

We don't need to fear the OECD countries so much, either. What we need to know is what is important to them and what we can use to negotiate strongly with them, and then try to apply this knowledge. That may happen in different forums, in different sectors, and in different regions for different purposes, but we need to keep that global monitor.

And, furthermore, having been a minister of planning for eight years in my own country, I'm very much aware of the weaknesses of planning. How do you connect planning with day-to-day reality? What happens many times — and this is part of the problem with the global approach — is that we are planning for a world that we want but that does not correspond to reality. Problems have to be solved, and solutions are, in effect, found by other means. Planning often becomes irrelevant. Somehow we have to make the planning which takes place for the global negotiations more relevant to actual situations, and that's why I suggest some prioritization. We need to begin to produce results so that the process will have credibility.

My final comment also relates to an aspect of planning, even though it may be changing. The mentality of planning was for twenty-five years attached to the notion of regulating. It ignored the forces of the marketplace and all kinds of other things. More modern planning recognizes the existence of some of those forces that need to be taken into account. We can perhaps bend these forces in this direction or that direction, but we shouldn't ignore them or stop them from functioning. A good deal of worldwide planning implied the regulation of all kinds of things. We should establish rules of the game in order to channel forces creatively but not stop them from serving a useful purpose.

Chairman Sewell: As a bibliographic footnote to this discussion, I commend to everyone the Interfutures Report of the OECD. It may not be accurate in its predictions, but it's very provocative concerning the implications of regionalism and of the linkages between various parts of the developing world and various parts of the industrial world under various presumptions about economic growth. The winners and losers don't sort out according to the normally expected pattern.

Now, Professor Weintraub.

Sidney Weintraub: I have a feeling that the people who talk about globalism have in mind a sort of gut reaction which equates regionalism with Northern forms of control and influence. They have in mind regional rather than general preferential trading arrangements. This is what Prebisch had in mind when he wrote his piece back in the 1960s.

But take food, for example. Food is a country issue, a regional issue, and a global issue. You can discuss food on all levels. It's silly to say you must only discuss food in all places in some global forum.

You can discuss energy on all levels. I think you can discuss most issues on all levels. You can even discuss trade on all levels, once you've got a satisfactory framework.

Even in their heyday, the United States aid programs had both regional and global aspects.

In other words, I really don't think there is very much to the issue except where preferential spheres of influence and all the ensuing gut reactions come up. Once you get into broader areas, I really think you're making much out of nothing.

On the architecture of international institutions, again, I think you are all trying to be much too neat, whereas architecture is eclectic.

There is a recommendation in the Brandt Commission report that the GATT and UNCTAD, because they deal in the same areas, ought to be combined as one institution. It's a fairly silly recommendation because it completely ignores all of the pressures and forces which created the two institutions.

I don't know how many agricultural institutions we have in the world structure now, but there are a good many, and I don't think you would want to combine them.

In your own countries, many institutions are probably doing comparable things, and every now and then there is an effort aimed at neatness and then it dissipates. I don't think institutional neatness is really an important issue.

Let me take the main functional issues of trade and money. What if we did reach an agreement, which Gerry called short-term, on a safeguard clause that was satisfactory? Wouldn't that have implications outside of the trade area? It would have tremendous implications in the industrial area. If it were the right kind of agreement, it might stop some of this talk about industrial policy whereby we reach orderly marketing agreements.

In the monetary system, what if we reached some kind of a short-term agreement whereby sufficient liquidity was provided for developing countries? Would that be the end of it or would it extend beyond that particular point into many of the other areas?

The point that Donald Mills made and Walt Rostow picked up — that we should not accept such agreements until we can reach an agreement on everything — is frightening because it implies that we should not make progress in any area until we make progress in everything.

I hope that Don didn't mean this and that I have misunderstood. But if that's a correct interpretation, it's dreadful, really. If it is correct, he has been giving very bad advice to the Group of 77. I hope I'm wrong on this.

I have one final point on discontinuities, legitimacy, and outcomes. There will be discontinuities in the world, and the real issue is not whether the world will continue as it has in the past. The world is always discontinuous, and I don't know what the major discontinuities of the 1980s will be any more than anyone else. Perhaps they will be in the area of debt and protectionism, and perhaps not. When we do meet these discontinuities, we need to ask how to deal with them and what kind of machinery we have for dealing with them rather than to ask whether there will be discontinuities.

And this brings me to the point of legitimacy. In all of my years as a government official, and this may be a terrible admission, I never thought of the word "legitimacy." To me, the concept means that when you confront critical problems, the focus has to be on production and employment. If the United States is stagnating, the first pressure is to grow.

I would have thought that the legitimacy of Brazil and Mexico and Jamaica is acquired not by playing a role in the discussions in New York, but by growing internally with a certain amount of justice to their populations. My thoughts have always been on issues like growth, employment, and distribution within countries. And it's that kind of issue which is the crucial one, not whether there is legitimacy.

So legitimacy and power may come by being part of a collectivity, but essentially it is demonstrated by various kinds of accomplishments at home.

Chairman Sewell: Professor Rostow.

Walt Rostow: If you had had, as I have, the privilege of sharing with eight Latin American colleagues and one Caribbean colleague a recent review of potentialities for cooperation in the Western Hemisphere, the notion that the United States could exercise any sort of hegemony in Latin America would not have occurred to you. There is a very strong assertive generation rising in Latin America and building up intra-Latin American institutions, both informally and formally.

So the real question is whether we can do something useful in the Western Hemisphere, or whether we should continue the enunciated policy of the Carter administration, which was: "Let's get the Latin Americans off our backs by taking them at their word and letting them go and get what they want in the global setting." That was the policy, aside from passing out Good Housekeeping awards for approved stances on human rights. It was not a very satisfactory hemispheric policy.

Roger Hansen raised a very important and interesting point concerning the relationship of regionalism to small and large countries.

Through an accident of history, I happen to have been involved in various aspects of regionalism since 1946 — almost without a break. And the truth is this: a small country is most vulnerable in a bilateral relationship. Its next most vulnerable position is a global relationship. Regional organizations are the natural friend of small countries.

Regionalism had its origins in Europe, where it started as a way of protecting the small and medium-sized countries against a revived Germany. To make it concrete, when you came to allocate coal and coking coal, which were then rarer than dollars in Europe, how did the smaller countries get their allocation? Once you set up a regional organization, there had to be objective rules. It wasn't a raw bilateral deal where power would prevail. This same approach and result can apply to the Western Hemisphere.

I might say the same about the Pacific Basin, which is not a revival of a Japanese "coprosperity sphere." It's sponsored by Australia and other countries with great care to make sure that the ASEAN grouping will not be overwhelmed, but strengthened. It would involve Canada and the United States, if it came alive, as well as the other countries of the region, and, indeed, some of the Latin American countries might wish to join it.

So the real question isn't whether this kind of hegemony is going to be revived. I think it's past history.

The relationship of certain African states, which are less developed, means that the Lomé type of agreement may be appropriate there. It is certainly not appropriate in Latin America.

The real question is whether you can do certain things more efficiently at the regional level, but let's not raise boogeymen.

Chairman Sewell: Don Mills.

Donald Mills: Let me first respond to Sidney Weintraub. The fact that developing countries have put forward an elaborate set of proposals does not by any means imply that we want no agreement on anything until we have agreement on everything.

I would pose the proposition in the very opposite way. If the developed countries as a whole could find it possible to agree on even a few of the many proposals that we have put forth — not in the last seven years but in the last twenty years, and specifically proposals dealing with structural change — then we would have some basis for developing a constructive dialogue.

It is the failure to get agreement on almost anything that makes it look as if we in the South are trying to get agreement on everything before we move. I believe even a very few agreements would have a tremendous psychological political effect on all of us.

Chairman Sewell: Gerry Helleiner.

Gerald Helleiner: I understood the proponents of the regional approach to mean that there is about to be a marginal shift in the North from multilateralism to bilateralism, and within multilateralism from globalism to more regionalism. And if it is only to be a marginal sort of change, I don't disagree with most of what was said. It is, I think, a matter of some concern if the changes in emphasis become more than marginal.

My concern is not with Brazil or Mexico. I don't think there is a single Latin American country on the list of least developed countries. My experience, unlike Professor Rostow's, is in tropical Africa, and my equivalent experience was serving on a Tanzanian delegation negotiating in the pre-Lomé period. I can assure you that the Lomé convention suggests how the developing countries can be short-changed in a regional agreement when you have fifty-six countries with no bargaining capacity whatsoever. They gain from global approaches. I'm not worried about the Latins.

Chairman Sewell: Nicolas Barletta.

Nicolas Barletta: I'm glad we have clarified the regional issue. And I want to repeat that it's very crucial that the issue not be considered a global versus regional one or a global versus sectoral one. The important thing is to make sure that we have the different forums where problem-solving can take place.

Chairman Sewell: In summing up, I want to return to my three points about the process, substance, and perceptions of the global agenda for negotiations. There is a vast, rich opportunity for creativity in melding global and subglobal approaches and for a kind of creative regionalism as we go into the 1980s. I happen to think that the area of the sub-Sahara and Africa is a development problem, separable even from South Asia for a variety of factors. This area is a pressing problem in terms of human needs, even if it is not the area containing the largest number of the poorest people.

So there is a great need to combine global and subglobal approaches, to use the political scientists' jargon. A lot of interesting developments will probably take place in the 1980s in terms of disaggregation within the industrial world. I'm not sure that, if you were to hold this conference ten years hence, many OECD shared interests would be recognizable any longer. That's going to be one of the interesting developments of the 1980s.

On the issue of substance, it seems to me there is a really remarkable convergence in points of view. I didn't hear anybody differ on the agenda issues in broad terms. Of the seven or eight points that Nicolas Barletta raised, four or five were listed by Jorge Navarrete as tentative focal points for the Cancún meeting. They encompassed a whole range of issues that we have discussed today.

There is one, however, that was missing, and from the agenda of this meeting as well. Let me just read a question from the audience which we are not going to be able to answer, but which eloquently states the issue that is not on the global agenda — using that word in a generalized sense, rather than to mean the agenda of any particular meeting, let alone the one at Cancún.

> I'm glad that the East-West issue was introduced. First, how effective can global negotiations be without the communist countries in the process? Their absence is another form of pressure on the South, indicating that socialism is not an internal, political, or economic option open to them.
>
> Second, the present trend of military competition between the United States and the Soviet Union will continue to increase and will divert resources as well as priority thinking. If there was at least one arena in which those two powers were on the same side, as in the North-South dialogue, surely military expenditures could begin to slow down. What a boost to new international lending institutions (as suggested by Mahbub ul Haq) there could be from this redirection of military expenditures by the major super powers.

This issue, which was eloquently addressed in the Brandt Commission report, remains on the global agenda in a broad sense.

Finally, I return to the issue of perceptions. The discussions of these two days illustrate quite clearly that perceptions differ, even among thirteen or fourteen people who have been engaged in these issues for a number of years and have met both officially and informally for a number of different discussions before.

And I return to Professor Rostow's point. It seems to me that progress will require a consensus, no matter whether that consensus is formally or informally developed. I would hope that consensus would come from a growing international understanding that the cost of inaction is going to be much higher than the cost of action, even though action is sometimes going to be costly.

The administration in this country, or in any other country, must see that, as important as East-West issues are, the Soviet Union could disappear tomorrow and we would still face a series of global food, energy, reindustrialization, and investment problems. These are probably going to be more important to our security and to global security than those issues directly concerning the Soviet Union. Until we acquire that perception, the dialogue will suffer in a variety of ways.

Elspeth Rostow: During this symposium we have spent twelve hours with a Group of Thirteen. Each of them said at one point, "I shall be brief." None was.

Aside from that, they demonstrated in microcosm the topic that we are discussing, breaking down fairly neatly into regionalists and globalists. In the end, with great courtesy, they agreed to differ. Even at the lowest moment of the interchange I felt as Mark Twain did about the music of Wagner: "Remember, it is better than it sounds."

If the purpose of this exercise is a continuing dialogue, we have now the material for just such a discussion. If we can be mindful of the issues raised here and realize that none of them is simple, that there is apt to be sharp controversy on each, but that their importance is generally agreed upon, then this session will have been amply rewarding.

Again, may I thank all the participants and especially Roger Hansen, who organized it all.

Postscript
Toward North-South Accommodation in the
1980s: Waiting for "Cognitive Evolution"

by Roger D. Hansen

We who care about these matters and have committed a large part of our lives to them must not merely echo in either polite or noisy or amusing ways the sterile nonsense that's been going on, with the South making its complaints to the North and the North managing to fend them off more or less skillfully.

If we can't come to an honest working consensus of the kind we had on foreign aid in Latin America and India in the 1950s — which embraced intellectuals and people who care in the North and the South — then the politicians won't listen to us in the moment of crisis.

Walt Rostow

We all thought [the Seventh Special Session in 1975] was a success because we talked mildly to each other, and we reached a "consensus" decision at the end. . . . In retrospect, it was really a failure, a conspiracy of softness. We should understand how and why this happened, but we have not learned the lesson, and therefore we have continued to delude ourselves.

We have not reached a broad consensus on the need for change in more than seven years of discussing the issue. There has been talk of consensus. But there is no consensus.

Donald Mills

I

The symposium presentations and the participant interchanges that followed them illuminated many salient perceptions of the present North-South conflict which are likely to have considerable influence on its evolution in the 1980s. Most are underscored in the opening and closing remarks made by the chairman of each of the symposium's four sessions. No further attempt at summarizing the proceedings will be made here. Rather, this concluding essay analyzes and critiques some of the most important perceptions which surfaced, and examines what they portend for the medium-term future of the North-South conflict.

The first set of perceptions meriting analysis encompasses those which lend strong and continuing support to developing countries' insistence on the New International Economic Order as a symbolic touchstone in the North-South dialogue. Despite their widely critiqued analytical deficiencies and their demonstrated capacity to evoke Northern responses which deadlock all serious negotiations, the proposals (and, more importantly, the *concept*) which together comprise the NIEO as a serious negotiating package continue to be supported by the South. What perceptions produce this tenacity, and with what implications for the present impasse?

Second, an equally persistent Northern perception deserves the same attention. It is the perception that the NIEO — as a symbol of a serious issue relating to "equity," "fairness," or "legitimacy" — is superficial, irritatingly inappropriate, detrimental to serious "mutual interest" negotiations, and readily dismissable if only the developing countries would correctly reevaluate their own self-interest.

Third, and related to this perception that was echoed throughout the symposium, is another Northern perception: that the so-called North-South dialogue over the NIEO and the set of "structural reforms" informing it can be dismissed without undue concern since "progress" in North-South relations is constantly being made in bilateral, subregional, and regional diplomatic settings where the focus is most often on individual "functional" issues. This particular perception was, unfortunately, not given the attention and analysis in the symposium which it merits.

Finally, greater emphasis should be given to the clashing of values being brought to the North-South dialogue. This hurdle to successful negotiations was perhaps most interestingly illustrated during the symposium in the conflicting concerns regarding the very purpose of economic development. Most simply stated, it was seen in the clash between those participants who gave highest priority to raising the standards of living of the poorest people of the developing countries and those who were most concerned with rejuvenating aggregate indicators of economic growth (GNP growth rates, rates of industrial growth, etc.). Whereas the former group of participants always probed *distributional* questions first, the latter group gave them little, if any, explicit attention.

Hopefully a brief analysis and critique of these crucial perceptions and values, so observable in the dynamics of the symposium, will permit some brief and informed speculation about the future of the present North-South impasse.

II

The first perception to be examined — that accounting for a great part of Southern (and Northern) support for the NIEO as symbolic of Southern demands — was underscored by the very first speaker, Mahbub ul Haq. He demonstrated its importance by literally casting aside his prepared remarks in order to answer Walt Rostow's observation, made the evening before the symposium began, that the NIEO negotiations "have been based on the wrong intellectual conception, the wrong agenda, the wrong negotiating forum, and the wrong cast of negotiators." In responding to Rostow's comment, Haq grouped his views around three central themes: the intellectual/economic rationale for the NIEO; the political rationale; and the rationale flowing from ever-greater degrees of North-South interdependence. Let us examine these themes sequentially.

Starting with the intellectual/economic rationale, Haq presented a view held by a vast majority of critics of the present international economic system, Southern and Northern. Supported by many other participants, he argued that the system is deeply flawed with significant imperfections and, perhaps more important, many anti-Southern biases. International limitations on the free flow of labor, capital, and goods are but the most obvious. Other major imperfections include the skewed nature of control over information and technology. The degree of oligopoly found in leading Northern export industries further discriminates against developing countries. Finally, Northern tariff and nontariff barriers to imports continue to be higher for typical Southern exports than typical Northern exports, and Southern agricultural exports are restrained because the GATT never succeeded in lowering barriers on agricultural goods to any significant degree.

Furthermore, as both Haq and Gerald Helleiner noted, the imperfections listed above appear to be on the increase, and significant concern was expressed regarding the "new protectionism" in the 1980s. The increased use of such trade-restricting practices as Orderly Marketing Arrangements, "voluntary" export restraints, and price-triggering arrangements (e.g., the steel-import program introduced during the Carter administration) are but a few of the more visible ingredients of the "new protectionism." These, then, are some of the major market imperfections and anti-Southern distortions in the present norms and rules of international economic relations that form the fundamental Southern critique of the present economic system and underlie the symbolic demand for the NIEO.[1] And any impartial examination of existing market imperfections and biases must conclude that the critique is for the most part valid. In this sense, Rostow's comment questioning the intellectual rationale of the NIEO was certainly too glib; if, on the other hand, Rostow was indirectly referring to the *content* of the NIEO proposals themselves, many of their own analytical weaknesses strengthen his provocative characterization.[2]

Several exchanges and comments during the symposium suggest how fundamental this set of perceptions is to the continuing North-South conflict. Those sharing these perceptions incessantly criticize the North for evading what are to them the most significant requirements of international economic reform — alterations which will rid the system of as many serious imperfections and anti-Southern biases as possible. The reforms suggested by these critics often go beyond changes in rules to alterations in institutions and voting procedures (i.e., "structural reforms," in the current jargon).[3] These structural reforms are seen as necessary to enforce a new system of norms and rules which introduce fairer and more equitable international economic "rules of the game."

On the other hand, the very vociferousness of the attack on so many aspects of the present economic system very often leads Northerners to counter with a sweeping defense of the system. Several interchanges between Gerald Helleiner and Sidney Weintraub are suggestive of the dynamic which can develop. In two instances Weintraub rightly insisted that we recall the successes of the developing countries in promoting economic growth over the past two decades. He reminded us that, in the aggregate, developing countries' growth rates during the 1960s and 1970s far surpassed those of the developed countries at similar stages in their own development. And while he did not demonstrate that the present international economic system was in any way causally linked to this rather impressive aggregate economic performance, his remarks clearly suggest the view that *at minimum* that system created a benign environment for the process of economic development. The essence of this Weintraub interpretation is shared by all knowledgeable defenders of the system. Judged by any historical standards, it has not only "worked," but it apparently has worked better than any previous system; and Southern countries, following appropriate development strategies, have been able to benefit significantly from the conditions created by it. As Joan Spero noted from first-hand observation, it is this "the-system-works" perception which pervades the thinking of most Northern negotiators in the present North-South dialogue, and which adds very significantly to their resistance to NIEO-type demands for sweeping changes.

To the detriment of any progress in overcoming the present conflict, the clash between these two sets of perceptions has not yet led negotiators and/or those managing the negotiations to reevaluate their strongly held views. They have been content to stand fast in support of their initial positions — positions reflecting sharply differing perspectives on the same empirical evidence. As the symposium discussion suggests, at least some of the difference can be explained by differing standards of judgment, which in turn often reflect a differing ordering of values. A pragmatist's standard of judgment produces the question: Does the system work? Often measuring by historical standards and aggregate results, the pragmatist answers that the system has worked well. This is the question and answer pattern which characterizes most Northern statesmen and economic policymakers. For most Southerners, both the question and the answer are quite different. The questions are: Does the system contain nonrandom biases against the South? Can it be made to work better? Is it showing signs of weakening in the 1980s, at the very time when greatly increased energy prices and Northern stagflation are causing severe stress in most developing countries? And the answers to each of these questions by typical Southern analysts (and many of their Northern counterparts, as the symposium contributions by Ed Hamilton, Gerald Helleiner, and Paul Streeten suggest) is yes.

One final point might be added to explain the crucial differences in perspective. We are analyzing a global economy in which the income gap between the developed and the developing countries in the aggregate is approximately ten to one. Furthermore, that relative gap — again in the aggregate — is *not narrowing*; indeed, in recent years evidence suggests that it may be increasing. And most important for some analysts, the income gaps between the people in the poorest of the developing countries and those in the rest of the world continue to grow, with no indication that this particular trend will alter. These income gaps between developed and developing countries naturally influence the perceptions about the adequacy of the present international system. Particularly among political elites and government bureaucrats, Northern "pragmatists" ask questions about aggregate global *growth*, and pronounce the system healthy. Southern critics ask questions about North-South income *gaps* and trends relating to those gaps, and label the system highly deficient. Each perspective accurately describes different trends and data influenced by the international economy and its present institutions, norms, and rules.[4]

As various symposium exchanges suggest, the initial plausibility of each of these sharply conflicting perspectives does much to enhance the current stalemate. In this sense, the symposium on several occasions reflected the interactive dynamics which are generally observed in official North-South negotiations.

An analysis of the Southern *political* rationale for supporting the NIEO as a symbol of Southern demands is equally revealing of a fundamental hurdle to North-South accommodation in the 1980s. As Haq noted in his opening remarks in the first session, the fundamental Southern goal in the political arena is what he simply called "a sharing of power." And he underscored the view that many Southern statesmen might even give this goal a higher priority than that of economic reform to increase Southern growth rates and shares of global production. For this reason, the NIEO is at least as much a political as an economic manifesto, and the calls for "structural reform" are calls for a greater Southern political voice in the shaping and management of international economic and political institutions.

Given Southern views that present international economic institutions like the IMF and the GATT and the political/economic processes governed by them are deeply biased against the South, the developing countries have attempted to increase their "voice" in these arenas in three ways. First, they have pressed for specific reforms of such specialized agencies of the Bretton Woods system as the IMF and the World Bank. The symposium reviewed many such proposals while adding to the list (see, for example, Paul Streeten's discussion of the IMF and Mahbub ul Haq's discussion of the World Bank in Session III). Second, the developing countries have attempted to use their voting power in the UN General Assembly and in special sessions of the UN to gain a greater degree of control over these specialized agencies, which are governed by a weighted voting procedure that leaves effective control in Northern hands. Finally, developing countries have created new organizations within the UN structure in which they have assured themselves of an equal or superior "voice." Examples include UNCTAD and the United Nations Industrial Development Organization (UNIDO).

The problem for the South is that none of these strategies has worked very well. Northern states, which control the specialized agencies through weighted voting procedures, have responded to a

certain degree (e.g., see Haq's review of the evolution of the World
Bank in Session III) to programmatic requests of the South, but not
to demands which would alter political control over the agencies
themselves. As regards the UN General Assembly, its votes are
nonbinding and do not control the actions of Northern states (even if
they do occasionally influence them). Finally, new UN organs and
agencies created by the South can also only recommend; furthermore, to the extent that their programmatic efforts require funding,
they must make do with whatever amounts Northern governments
decide to contribute to their activities.

The typical Northern response to the Southern demands for
power-sharing was well represented at the symposium. Sidney Weintraub, who opened Session II, noted that in international relations
power is not granted, it is (somehow) won. And when it is won, then
and only then are international "agendas" altered and negotiating
outcomes changed. Walt Rostow made the same point on several
occasions, specifically linking "power" in international relations
with the process of rapid economic growth. And again, like Weintraub, Rostow contended that negotiating outcomes *reflect* a preex-
isting distribution of global power; they do not *alter* that distribution
of power and political influence. This perspective on the issue of
power and political "voice" in international relations may help in
great part to explain Weintraub's disinterest in the issue of "legitimacy" in international economic relations and Rostow's concentration on the "short run" as opposed to the "long run" in the symposium's analysis of international economic problems and institutions.
For if international institutions like the IMF and the GATT simply
reflect the goals and values of the world's most powerful states, if
regimes like the present trade and monetary regimes are altered only
in response to changing power relationships, and finally, if "power"
is never deliberately shared by "status quo" states but only ceded in
response to demands made by (relatively) strengthened revisionist
states, then issues of "legitimacy" and the "long run" — at least at
first glance — appear to lose most of their relevance to the pragmati-
cally oriented statesman, bureaucrat, and/or analyst.

To summarize: the South has invested in the symbol and content of
the NIEO demands for "shared power" to alter international institutions, which are thought to assist in skewing the distribution of
global income and political influence in favor of the North. Most
Northerners involved in NIEO analysis or negotiations view
"power" as something which is aggregated through processes like
rapid economic growth, and which is a *prerequisite* to the changing

of international institutions, regimes, and rules, such as those in the fields of trade and monetary relations. Regimes, their norms and rules, *can and will change in negotiations only in response to altered power relationships.* Ergo, according to most Northern observers, global negotiations which have as a *goal* the altering of power relationships are useless; for *unless a power shift has already taken place, stronger status quo powers will concede nothing to weaker revisionist states.*

It is interesting to note that this (hopefully fair) representation of several "Northern voices" in the symposium is completely congruent with "mainstream" United States (if not most Northern) scholarship of political scientists and particularly international relations analysts. Subject, of course, to the usual number of caveats about leads and lags in state behavior — for reasons concerning which there is not as yet any consensus — most members of the international relations profession within the United States would accept the following characterization: when the distribution of power among states changes, individual state behavior and the outcomes of international negotiations will change. And with regard to the issues and arenas of state interaction discussed in this symposium — which political scientists would currently refer to as a broad range of economic "regimes," including those covering trade, monetary affairs, energy, and many others — there would be a considerable consensus that such "regimes," with their norms and rules, are particularly dependent upon the power and perceived interests of a single hegemonic state or a small group of powerful states which conceived and nurtured those regimes. As one commentator has recently noted with relation to such international economic regime questions, "underlying power capabilities are critical for the establishment of a new system, and they determine whether that system can be maintained if it comes under attack."[5] In analyzing the present North-South impasse the same commentator makes two further points which relate quite specifically to the discussions of "power" and the political desires of the developing countries as expressed in the NIEO. First, international economic regimes are likely to be stable only when they operate in ways "that benefit those states that have the most power"; "a system is likely to be unstable when there is a lack of congruence between underlying power capabilities and actual practices . . ." And second, if the developing countries lack "the underlying power capabilities to sustain their vision of a new regime, such a regime is bound to come under persistent attack."[6] It is not surprising, therefore, to find this international relations specialist making arguments similar to those of Rostow and Weintraub: since the South does not have the power to force the changes it desires, and since the North will not share that power or alter those institutions which are thought to maximize their own interests, the South should turn away from *inevitably fruitless global negotiations over power-sharing* in a new economic order and concentrate on opportunities for immediate, pragmatic economic gains which may be produced in issue-specific bilateral and regional negotiations.

On first glance, the clashing of Northern and Southern perceptions on the "sharing of power" goal implicit in the NIEO demands would indeed seem to doom all prospects for accommodation at this level of the North-South dialogue. And this is quite possibly why Walt Rostow's Session IV gesture of compromise to Donald Mills was the suggestion that the NIEO "dialogue" continue at the "global level" while "practical efforts" to make constructive progress should take place in regional settings and focus on the specific functional issues such as energy, food production, and the environment, which Rostow identified in his Session I presentation.

The more one examines the set of issues illuminated by this debate over "power" and its relation to the negotiating problem raised by the NIEO as symbol, however, the more one is impressed by the indeterminancy, irrelevance, or fundamental weakness of a great many of the "classical realist" or "neorealist" postulates concerning the behavior of nation-states in the final quarter of the twentieth century.[7] While there is not enough space in this postscript to examine in any detail the many complex issues that obviously require lengthy and thoughtful analysis, a partial listing of them in brief form may help to indicate that much of what presently passes for conventional wisdom shouldn't. The following discussion should suggest that (1) the concept of "power" as generally used in the analysis of international relations today is often so ill-defined as to be useless at best and misleading at worst; (2) present trends in international relations made the definitional effort *increasingly necessary* — as almost all "neorealists" recognize — since the more attention we pay to the problems of concept and definition, the more aware we become of the contextual (specific, real-world) constraints on the concept, and on the utility of "power" itself; and, finally, relating specifically to the North-South conflict, (3) the rather simple notion that, without significant increases in "power," the South will be unable to engage the North in a serious negotiation concerning many of the symbolic political aspects of the NIEO misrepresents the range of possible outcomes. In short, it "overdetermines" the outcome of a highly complex confrontation.

First, with regard to the conceptualization and specification of power, almost all international relations specialists readily admit problems of measurement. Having done so, many proceed to forget the problem and use the concept in undefined and overaggregated ways which seriously undermine the effectiveness of their arguments.[8] Others construct indices of power which serve only to further illustrate the problem of measurement. To take but two examples, authors currently attempt to measure relative "economic power" by comparing aggregate GNP figures of states, and attempt to measure "international financial power" by comparing total foreign exchange reserves.

It doesn't require much probing analysis to reveal the obvious limitations of such measurements. With regard to aggregate GNP, a country may rank very high on such a list, but its *dependence* upon external sources for crucial inputs to achieve and maintain its GNP "power" may be either minimal or very great. This obvious point was made several years ago in a more general way by Kenneth Waltz in comparing the potential effects of the OPEC actions of 1973–74 on Europe and Japan on the one hand (which were very dependent on OPEC oil) and the United States on the other (which was, at that time, relatively independent of OPEC oil).

The same ambiguity surrounds the use of foreign exchange as a measure of international financial "power." Take the case of Saudi Arabia. Does the explosive growth in Saudi international reserves really provide an adequate measurement of "power" in the international financial arena? More votes in the IMF, yes. More consultation, yes. But if power retains anything like its standard meaning of the capacity of state *A* to influence the action of state *B* with respect to issue *C*, do those reserves increase Saudi "power?" Or does their very size increase Saudi *dependence* on the health and stability of (Western-dominated) international financial markets? The *value* of those reserves is to a large degree a function of the strength of the dollar at the present time. Therefore, what some would choose to emphasize as a "power capability" now in the hands of Saudi Arabia, others would view as a growing vulnerability. The truth is probably somewhere in between. For present purposes, all that need be established is that in the present setting of international relations, "power capabilities" are never as unambiguous as they seem to many analysts of world politics.

One further crucial issue must be raised with regard to the conceptualization and definition of power, an issue which also serves as a bridge to the second set of observations. Do we accept the relational definition of power in international relations which is common to other social science disciplines (i.e., the capacity of A to influence B with respect to issue C)?[9] While an overwhelming number of "neorealists" and "classical realists" appear to do so, a leading realist, Kenneth Waltz, does not. In a passage which does little for the cause of definitional clarity but much for thinking about North-South relations in the coming decade, Waltz writes:

> Whether A, in applying its capabilities, gains the wanted compliance of B depends on A's capabilities and strategy, on B's capabilites and counterstrategy, and on all these factors as they are affected by the situation at hand. Power is *one* cause among others, from which it cannot be isolated. The common relational definition of power omits consideration of how acts and relations are affected by the structure of action. To measure power by compliance rules unintended effects out of consideration, and that takes much of the politics out of politics.[10]

Waltz then offers a specific definition of his own: "the old and simple notion that an agent is powerful to the extent that he affects others more than they affect him."[11] Adding to this definition, which moves increasingly toward the concept of autonomy, Waltz notes that "[to] be politically pertinent, power must be defined in terms of the distribution of *capabilities*; the extent of one's *power cannot be inferred from the results* one may or may not get."[12]

One of the most obvious problems with Waltz's attempted definition (or conceptualization) of power is that it simply throws the burden of definition from "power" to "power capabilities." What is a power capability? As Baldwin and others insist, we cannot possibly identify a power capability unless we specify "who is influencing whom with respect to what; in short, both scope and domain must be specified or implied."[13] In how many different policy-contingency contexts can it be used (i.e., how "fungible" is it)? Baldwin adds: "What functions as a power resource in one policy-contingency framework may be irrelevant in another. The only way to determine whether something is a power resource or not is to place it in the context of a real or hypothetical policy-contingency framework."[14] Thus, except for what may follow from the implicit Waltzian equation of power with autonomy (low degrees of vulnerability, low degrees of *interdependence* when the concept is appropriately defined as *mutual vulnerability*, and, at the extreme, a capacity for autarchy), his definition of power does not escape the dilemmas of the generally accepted relational definition.

The second set of observations relates to the following point: present trends in international relations make this definitional effort increasingly necessary, as all neorealists realize. Their analysis in recent years has led them to adopt several perspectives which distinguish them from their "classical realist" progenitors. One of the most important new perspectives relates to the multidimensional nature of power.[15] Power is no longer viewed as readily fungible. "Power capabilities" in one arena of international conflict may be quite irrelevant in another. Therefore, power analysis is now moving to specific "issue areas," e.g., trade relations, financial/monetary relations, "international common areas" (i.e., the oceans, the Antarctic, global environmental problems, etc.). Analysis of traditional realpolitik concerns (military/strategic issues) continues. What has been dropped is any *assumption* that "power capabilities" in the military sense, once felt to be ultimately relevant to almost all conflicts, can indeed be "used" in disaggregated, issue-specific segments of international relations.[16]

A second "neorealist" perspective worth noting relates to the notion that growing degrees of international interdependence increasingly limit the consideration (let alone the use) of more traditional "power capabilities" of the military type and increasingly lead states to search for and negotiate mutually acceptable "rules of the game" (in the present jargon, "regimes") in arenas of interdependence. In this process, states will attempt to use whatever issue-specific "power" they possess, but once again the complexity of the power issue and the "translation problem" (that of converting favorable "power capabilities" into favorable "bargaining outcomes") arises. Perhaps this process can best be illustrated in a manner relevant to the North-South arena by the dynamics of coalition formation and issue linkage in the Law of the Sea Conference. By holding their coalition together through the G-77 institutional mechanisms and norms of behavior, and by insisting upon the so-called "deep seabed regime" in exchange for accepting Northern proposals regarding such issues as limits to territorial waters, passage through straits, and other proposals viewed as crucial by Northern states (for both security reasons and economic reasons), the developing countries were finally able to extract concessions from the North relating to deep seabed mining, resource transfers, technology transfers, and a new Southern-controlled international agency, concessions which one astute Northern commentator, writing during the negotiations, argued would never be granted.[17] Since the Reagan administration withdrew United States approval of the LOS III draft treaty at the last minute in order to "review the negotiation," the outcome is still in some doubt. Either way, the illustration suggests the tenuous relationship between apparent "power capabilities" and bargaining outcomes in an international environment increasingly characterized by (1) self-imposed limitations on the use of military force by "great powers"; (2) growing degrees of interdependence leading to negotiated attempts to provide required degrees of policy collaboration or coordination; and (3) the emerging "global problems" of the type mentioned in the symposium by Rostow and identified by international relations analysts for at least a decade.

This brings us to the final point to be considered with regard to power: the rather too simple notion that without relative increases in their "power capabilities" the South will be unable to engage the North in a serious negotiation regarding many of the emotionally charged political aspects of their demands symbolized in the NIEO call for structural reform. The preceding discussion of power and its growing ambiguities in the present international environment should by now suggest the overly mechanistic and overly deterministic manner in which this aspect of the North-South conflict is often analyzed. Power "capabilities" or "resources" will alter from arena to arena of the international politics of North-South relations, all the more so because in this slice of international politics Northern states will seldom — if ever — resort to military self-help measures; exceptions will most often occur where East-West and North-South fault lines coincide. If the nature of power capabilities will constantly be in question, so too will the capacity of Northern states to translate "capabilities" into control over outcomes. Indeed, it is one of the paradoxes of the study of international power that some of the most perceptive "neorealists" who are closest to their classical counterparts are also most acutely aware of (and anguished by?) growing domestic impediments to "realpolitik" behavior in advanced developed countries, particularly the United States.[18] For over two decades analysts of industrialized states have been pointing to the clash between rising domestic welfare demands and forceful, coherent, and effective foreign policies. The clash continues, and discounting cyclical variations, the trend would seem to be in the direction of less forceful, less coherent, less effective foreign policies as advanced industrial states are increasingly constrained by a rapidly multiplying number of domestic constituencies and demands, international forces of interdependence, and "weak" nation-state assertiveness, none of which the advanced industrial states can effectively control without more policy coordination than they have thus far been able to achieve. And while some political scientists predict a renewal of "strong states" in the North in response to these domestic and foreign challenges, others project a trend in which Northern state power is slowly effaced both from above (i.e., the challenges of the international political, economic, and social environment) and below (i.e., domestic constituency challenges).[19]

If the international agenda of issues for negotiation begins to tilt toward such global concerns as nuclear proliferation, environmental problems, food production, and population control, "first-best" solutions will require Southern cooperation. One "second-best" solution of exclusive Northern (or "Trilateral") cooperation in arenas where it seems technically feasible will always be proposed by those who despair of "global" cooperation, and continued North-South stalemate will encourage a turn to this form of the second-best.[20] But for all the reasons cited above, the 1980s would seem to be a decade in which the opportunity for something resembling North-South bargaining — at levels more comprehensive than the bilateral, regional, and functional — will be ever present. Whether these opportunities are embraced or rejected is another matter.

Little need be said about Haq's final rationale for the NIEO in response to the Rostow challenge: that is, the moral rationale which hinged on growing interdependence and Haq's view that this interdependence was finally legitimizing the "domestic analogy" in international relations. Rather than considering the moral implications of Haq's position, in light of the preceding analysis one final word might be ventured about the eventual impact of interdependence on the North-South stalemate.

Despite the fact that North-South relations are at present characterized far more by dependence than interdependence (i.e., *mutual* vulnerability), the steady evolution in the latter direction seems obvious and inevitable. Arguments that the North now needs the South in order to escape the stagflation trap are highly exaggerated. Yet even now Southern markets absorb more United States exports than any of our Northern trading partners, and represent the fastest growing United States foreign market. In 1973 the United States could (apparently) afford to face OPEC price increases and embargo actions with considerable impunity. By 1980 it could not stock its congressionally mandated oil reserves without Saudi Arabian consent. The United States can embargo food shipments to the USSR, but Argentina can then play a major role in filling the Soviet grain deficit.

The indicators are already obvious, and one previously noted element should greatly enhance North-South interdependence in the 1980s. It concerns the so-called "global agenda" of problems such as food production, environmental safety, protection of remaining "global commons" from ocean fisheries to radiowaves, nuclear proliferation, and a host of others. The increasing salience of such problems, resulting from the technological explosion which is causally linked to most of them, will surely not help to break the present North-South impasse unless both parties to the conflict reevaluate their own position and that of their rivals in light of these new challenges and our fuller understanding of them. The argument is simply that the costs to both sides of the stalemate should become increasingly obvious, heightening the probability of explorations for breakthroughs.

In retrospect, then, what presently appears to some symposium participants (and most international relations analysts) as a Southern "asking" for political power without the capacity to wrench it from Northern hands, and the utter futility of negotiating about "power-sharing" because "that's not the way states operate," should seem less and less utopian as and *if* Northerners seriously begin to analyze the concepts of power, power capabilities, interdependence, resource limitations, and "global agenda" problems facing the international system in the coming decade.

Indeed, many Northern leaders are already making somewhat similar calculations. The result has been the emergent strategy of cooptation, whereby leading Southern states possessing "real power" are invited to "join the club." Having analyzed the strategy and its glaring weaknesses elsewhere, I will not repeat the effort here.[21] What bears repeating is the obvious conclusion that issuing the invitation after "real power" is achieved is a bit too much like closing the barn door after the horses are already gone.

In sum, realpolitik behavior (don't "give away" any power you can still defend), which may have formed the essence of prudential statecraft in a different international setting, would seem to be inappropriate enough to the 1980s to merit and actually receive serious reexamination. Only such a process of "cognitive evolution" at many levels can possibly turn the present North-South conflict into a relationship which allows increasing (and increasingly needed) degrees of cooperation.

III

If Mahbub ul Haq's response to Walt Rostow's challenge to the NIEO revealed some crucial challenges to North-South cooperation, so too did Rostow's presentation in Session I of the symposium. From his perspective, the North-South negotiations that have taken place since 1974 are based on the wrong intellectual framework, the wrong agenda, the wrong negotiating forums, and the wrong cast of negotiators. As Rostow notes, this view emerges from his belief that by the end of 1972 the world economy had once again entered a cycle marked by shortages of and relatively rising prices for raw materials, led by food and energy, and that some additional problems such as global ecological damage were also changing the priority of the world's economic problems. These considerations lead Rostow to deplore the folly of continued NIEO negotiations and to emphasize the urgent need to alter agenda items, move negotiations from the global political level to the regional technocratic level, and focus on functional/sectoral problems rather than "global bargains."

Even if everything Rostow believes about the nature of global economic problems since 1972 is accurate, and even if his policy proposals are an appropriate response to those economic problems, his apparent lack of interest in or idiosyncratic reading of the history of the North-South conflict severely limits the effectiveness of his case *if* it is meant to overcome the present North-South impasse.

At the very moment when Rostow suggests that the nature of global economic problems was changing, the South as a bargaining unit in international relations and its commitment to a carefully negotiated and refined agenda reflecting all the compromises necessary to hold a coalition of over 115 countries together was reaching its highest plateau in the history of the movement. In September of 1973 the Nonaligned summit meeting in Algiers for the first time adopted almost to a word the economic agenda which had been produced by ten years of Southern cooperation and compromise within UNCTAD, where countries had worked together not as the Nonaligned countries, but as members of the so-called Group of 77. Thus the Algiers summit of 1973 witnessed a historical event — in effect, the adoption by both institutions of Southern cooperation of the same priorities, the same agenda, and the same strategy for achieving those goals. The final resolution of the Algiers summit went further; for the first time a call was issued for "a new international economic order," and plans were announced for a special session of the UN General Assembly to analyze ways in which that

organization could further assist in the process of economic development within the South. In these two separate but related elements of
the Algiers resolution, the Nonaligned countries coined the phrase
(NIEO) that will be with us for years to come and set in motion the
events leading directly to the Seventh Special Session of the UN
General Assembly of 1975 and to all the attempts to use that body to
institute the NIEO since that date.

It is not necessary to analyze here the range and depth of Southern
feelings which ultimately led to the unified commitment to an agenda
and a program of action in Algiers in September of 1973. It is enough
to recall that the South's economic complaints had a twenty-year
history that is well documented in UN debates and elsewhere, and
that the Southern demand for a new international agency to deal
with many of these complaints ultimately succeeded with the creation
of UNCTAD in 1964. UNCTAD was the birthplace not only of the
Group of 77, but also of the "formalized" North-South conflict.
The developing countries' frustrations with the negligible successes
achieved within UNCTAD in great part account for the economic
thrust of the Algiers summit of 1973.

The political grievances of the developing countries are equally
traceable through the background, founding, and slow development
of the Nonaligned Movement. From the Bandung conference of 1955
to the Fourth Summit of the heads of state of the Nonaligned
Movement in Algiers, the member countries were sorting out their
own complex differences and their strategy toward the countries of
the "Western alliance." The scars left by Western colonialism and
racial bias were never very far from the surface in these gatherings.
And if their original demands were political in nature, by 1973, as
noted, they had incorporated the economic demands of their "sister"
organization, the Group of 77, bag and baggage.

If the institutions, the communications networks, the shared
norms, values, and behavior, and the agenda of grievances were thus
prepared for an assault on "the system" of Bretton Woods by late
1973, the actions of the OPEC states, just weeks after the Algiers
summit, in raising oil prices and embargoing oil shipments to certain
Western countries provided the shot which opened the assault.

It is this rich and complex political, economic, and psychological
history which Rostow's detached and detailed economic study of the
"fifth Kondratieff upswing" ignores. As an economic historian,
Rostow has no particular reason to pay any attention to this history.
But as an advocate of a set of propositions relating to the present
North-South conflict, he must do so. For if he doesn't, there is the

constant danger that his own basic concerns will be disregarded and/or caricatured as simply one more example of American technocratic positivism at work.

This prospect is particularly likely when Rostow advocates a *regional* and a *functionally specific* approach to today's global problems and the North-South conflict. Both aspects of his proposal are broadly perceived in the South as Northern attempts to undermine thirty years of effort to achieve enough unity among developing countries to bargain with the North. From the Southern perspective, maintaining that unity vis-a-vis the North (regardless of marked and continuing South-South conflict in many geographic regions and functional arenas) requires large degrees of *globalism* and *functional issue linkage*. Both are needed in order to hold together intra-Southern compromises on various elements of their demands. *A priori*, therefore, regionalism and de-linking are highly suspect; they are generally viewed as Northern strategies to undermine the Southern challenge and to support the perpetuation of the present "inequitable" institutions and rules governing the global political economy. As Haq, Helleiner, Mills, and Navarrete noted in the symposium, the South certainly shares Rostow's "short-term" concerns and, therefore, is not opposed in principle to his agenda. The strong opposition arises the instant it appears that work on the "short-term" agenda might undermine efforts to achieve the South's "long-term" goals, which relate to altering the structure and functioning of the existing international economic system. As the reader of the symposium transcript will recall, the "short-term" versus "long-term" issue became one of the most heatedly unresolved points of the symposium discussions. It is an issue that brings us back once again to a division between those relatively satisfied with the status quo and willing to alter it marginally on a case-by-case basis, and those relatively dissatisfied with the status quo and determined to revise it substantially.

Thus, while there was a universally shared concern with the vital problems raised by Rostow, there was little but conflict over his policy prescriptions. Nevertheless, in various observations, most particularly those of Navarrete and Haq, there did emerge the outlines of a compromise which deserves very serious attention. As Navarrete noted in his exposition of the planning for the October 1981 summit to be held in Cancún, Mexico, each of the items presently being considered for discussion would be examined from both a short-term and a long-term perspective. For example, in addressing financial and monetary issues, the approach being considered calls for a focus on the short-term problems of developing countries' deficits and the issue of recycling and on the long-term issue of the transfer of resources to developing countries and the institutional aspects of that issue. Anyone familiar with the North-South conflict can see behind those code words a combination of the Rostow agenda, the NIEO agenda, and a very strong — if implicit — link between the two within a "functional area." Is there, in this approach, an opportunity for an escape from present negotiating patterns and results? Without a concomitant process of cognitive evolution, probably not. But if new data and new concerns of the type raised by Rostow can impact upon the negotiators' perceptions of the problems facing them and perceptions of their counterparts' desires, the probabilities may be much improved.[22]

It is from this perspective that one views the major flaw in United States policy toward the North-South dialogue in particular and Northern policy in general as the failure to develop a positive agenda of its own in response to the calls for a new international economic order. For eight years, the North has been satisfied with analyzing the flaws in Southern proposals and rejecting most of them out of hand. There has been no attempt to "upgrade" the ongoing negotiation with counterproposals rather than rejections. Most impartial analysts would agree that many proposals contained in the NIEO demands merit either outright rejection or substantial alteration. But having failed to involve itself in this process in a constructive manner since 1974, the North is still faced with demands presented at that time. That makes the task of altering the agenda now far more difficult than it might otherwise have been.

IV

Another Northern perception, echoed in the symposium as well as in the corridors of Washington and other Northern capitals, that is a constant contributor to continued stalemate is the following: the "dialogue" over the NIEO and the "structural reforms" contained in it (i.e., most of the political/institutional reform content) can be dismissed without undue concern since "progress" in North-South relations is made daily in bilateral and regional settings relating to a host of functionally specific issues. This perception was revealed on several occasions during the symposium. Joan Spero, for example, noted that it was the prevailing view in Washington while she was the United States ambassador to ECOSOC; Sidney Weintraub associated himself with the same view during Session I. Unfortunately, the subject was not analyzed in any detail in the symposium, but a word must be said about it here. For if this view prevails, it quite naturally runs counter to any serious attempt to develop further the combined "short-term"/"long-term" approach just discussed.

There is no denying the fact that "things get done" at the bilateral and regional levels on functionally specific issues. And there is no denying the fact, alluded to by both Spero and Weintraub, that Southern finance ministers don't seem much interested in pressing NIEO demands when talking with official Northern representatives in their own countries. But there is, of course, a quite natural explanation. Finance ministers in developing countries, like their counterparts in the North, are overwhelmingly concerned with important day-to-day problems. Indeed, for some Southern finance ministers, the main day-to-day question is how to keep their countries financially afloat. To press for NIEO "structural reform" with Northern officials who are often in a position to control a Southern country's access to both public and private capital markets would surely be quixotic in the extreme; quixotic finance ministers constitute a small universe.

Does it follow that, because Southern finance and trade officials in their own capitals generally adopt a pragmatic, short-term "problem-solving" attitude unaccompanied by NIEO rhetoric, commitment to the goals and symbols of the NIEO — again, oversimply, the political/institutional reform elements in it — is limited to a few Southern officials exiled to the UN because they have little influence at home? There does not appear to be much evidence to support this assumption. The Southern push for the proposals included in the NIEO demands takes place not simply at Turtle Bay. It is revealed at each UNCTAD meeting where ministerial-level Southern representation is the norm; at LOS III negotiations over the better part of a decade; at the Conference on International Economic Cooperation (CIEC) in Paris (1975-77); at Nonaligned summit meetings and many others. In these varied settings, a mixture of representatives — ambassadors, ministers, and heads of state — have met to set agendas and coordinate strategies, all with the shared goal of achieving the major political/institutional objectives enumerated or implied in the original NIEO document. The evidence thus suggests simply that Southern states have to survive in the world in which they find themselves while they continue in their attempts to change it.

The crucial question, then, relates to the opportunity costs to the North of continuing to avoid serious negotiations over the longer-term "structural" issues being pressed by the South. And this question brings us back to earlier considerations about the nature of international politics, power, interdependence, and "global agenda" problems in the 1980s. Analysts do not agree on the opportunity costs of the present Northern strategy of rejection plus attempts at cooptation. Some clearly judge the costs appropriate; others feel that they are already far too high and continually rising with the passage of time. But at least the notion should be dispelled that there are no costs involved because all the Southern talk in the UN is a charade of little importance to Southern governments. It simply does not withstand close scrutiny.

V

A final set of issues reflected in the symposium but never pulled together directly concerns the problem of assistance to the poorest people in the developing countries (often referred to as the "absolute poverty" or "basic human needs" problem). The issue was first raised by Paul Streeten in Session I, and was subsequently addressed from varying perspectives by Ed Hamilton, Mahbub ul Haq, Nicholas Barletta, and Walt Rostow. Their comments suggest the fundamental normative and practical problems involved in attempting to deal with the absolute poverty problem. They are brought together and underscored here because of the disaggregated manner in which they arose and are, therefore, reflected in a reading of the symposium transcript.[23]

In normative terms, the issue of what can and should be done — and by whom — to assist those one billion persons living on the razor's edge of existence is one of the thorniest in the North-South debate. As Haq pointed out, development experts are just beginning to understand how the vicious cycle of undernourishment, lack of health and educational facilities, and lack of remunerative employment opportunities can be circumvented in cost-effective and bureaucratically manageable ways. Haq also expressed his concern that at this crucial moment, the interests and resources of the World Bank, which has been up to now one of the leading institutions in illuminating, analyzing, and proposing programs to overcome this problem, may soon turn in other directions.

How much priority does this aspect of the development problem in the North-South dialogue merit? For Paul Streeten, the priority seemed almost absolute, as suggested by his remarks in Session I and

Session IV. Haq's perspective seemed very similar to that of Streeten's. Most other participants paid more than lip service to the problem, though a continuum of views was clearly represented. Perhaps the differences were most clearly shown in the exchange on the merits of regionalism that took place in Session IV. And this is logical since the great bulk of the world's absolute poverty population lives in one region of the world, South Asia.

From the pragmatic perspective, only one of the major existing issues was raised. Ed Hamilton underscored the necessary linkage between continuing Northern financial flows to assist in overcoming absolute poverty and demonstrable evidence that those aid flows are having their intended impact. Unexamined were these equally critical questions: (1) How should funds for this purpose be raised, allocated, and distributed? (2) What kinds of oversight mechanisms would be required to limit the potential problem implicit in the linkage noted by Hamilton? (3) What kind of "evidence" can demonstrate that programs in this area — many of which require long lead times — are working? (4) Can the evidence which Hamilton suggests will be necessary to assure continued Northern financial assistance be gathered without posing insuperable obstacles to Southern states' sensitivity concerning their national sovereignty? Fundamental as they are to both the normative and practical aspects of the present North-South problematique, an exchange centering on these issues will have to await another symposium unless, through some inexplicable circumstances, they actually find their way onto the official North-South agenda.[24]

VI

Thus far this retrospective on the symposium has been rather less than sanguine with respect to the existing potential for North-South accommodation in the coming decade. And yet it does not seem entirely implausible, based on the two-day interaction and its results, to conclude on a note of cautious optimism. First the optimism, and then the caution.

This optimism reflects what appeared to be a growing degree of convergence around (1) an appropriate agenda, (2) a plausible balance between the short-term problems critical to both sides and the long-term reforms of fundamental importance to the South, and (3) a *de facto* (or implicit) handling of the "linkage" problem which might be acceptable to both North and South.

Regarding all three issues, Jorge Navarrete's discussion of the planning process for the Cancún summit is revealing of the potential convergence of ideas and interests irrespective of the outcome of any single conference. With regard to an appropriate and "balanced" set of issues, the five elements thus far receiving major attention in preparation for Cancún do suggest room for considerable "mutual interest" bargaining: (1) food and agriculture; (2) trade; (3) energy; (4) financial and monetary issues; and (5) interrelationships between resource use in economic growth patterns, population growth, and environmental issues. Obviously, the scope of this gradually emerging framework encompasses issues of significance to both sides and represents a *de facto* melding of Walt Rostow's "new agenda" and the South's NIEO agenda.

If we look beneath the five points receiving priority consideration to date, there would also seem to be a purposive blending of short-term and long-term issues within each major category. In the food field, the balance is between short-term problems of food security systems and long-term problems of agricultural development to increase Southern degrees of agricultural self-sufficiency and ease the problems of rural poverty. In the trade field, there would seem to be an acceptable balance between the attention to be given to the immediate problems of "damage limitation" (of the type referred to by Gerald Helleiner in his Session III analysis of the "new protectionism") and that given to the longer-term issue of a coordinated approach to continued industrialization in the South and programs of "reindustrialization" in the North. This longer-term problem is already on Northern agendas, as several recent studies by the OECD and the Trilateral Commission testify. If this quieter approach can rid the negotiations of the explicit and acrimonious Southern demand of producing 25 percent of the world's manufactured output by the year 2000, a process may begin which moves in a direction desired by the South but in such a way as to merit and achieve Northern cooperation.

In the energy field, the same balance between short-term and long-term concerns exists. First comes the problem of limiting the present growth restraints caused by rapid energy price increases; second comes the effort to develop and finance appropriate long-term energy strategies for the most energy-deficient Southern countries.

As noted earlier, the emerging approach to the finance and monetary area likewise incorporates short-term and long-term considerations. The former focus on immediate problems of petrodollar recycling and the management of the current debt problem in the South; the latter focus on the longer-term problem of adequate resource transfers from North to South, including efforts to alter international and regional financial institutions to improve their capacity to assist in this arena. In this category, one is reminded of Haq's suggestion for strengthening the World Bank and Paul Streeten's acute observations about the present limitations on the International Monetary Fund's capacity to manage either the short-term or the long-term problems in this field, and the nature of needed institutional reforms. Both of these Session III contributions offer valuable insights and operative suggestions relating to the fourth item likely to receive significant attention at the Cancún summit.

Finally, the issue of linkage between these items. While it is as yet unclear how explicit the issue of linkage between different part of the Cancún discussions will become, a brief agenda and the short-term/long-term balance within each agenda category suggest that much thought is already being given to the linkage issue. It may well be that if the summit meeting leads to the breaking of the present impasse on the "Global Negotiation," the problem of progress in several major functional areas *simultaneously* can be agreed upon *de facto*, avoiding the type of clash over explicit issue linkage which has caused so many problems for North-South relations ever since the Paris CIEC meetings of 1975-77. The developing countries are highly unlikely to discard their commitment to issue linkage, since they fear that without it the North will simply move forward in the energy area and forget the rest. Rightly or wrongly, the Group of 77 has always felt that the capacity to link issues was one of the South's few real "power capabilities" at the present time. The value of the Cancún modus operandi would appear to be its willingness to leave certain items formally unaddressed while paying appropriate attention to them in the very manner in which issues are raised and in the timing of actions in different areas.

It should be noted that symposium remarks by Hamilton, Haq, Helleiner, and Streeten all converged around the notion that an approach resembling Navarrete's description of the Cancún summit plans might be an optimal way to balance Northern "system maintenance" concerns with Southern "system reform" goals. Haq's notion —that the best opportunity for melding Northern "problem-solving" perspectives with Southern "structural reform" goals may be presented when North and South jointly consider the reform of present institutions and/or the creation of new ones to manage short-term problems

— is intuitively appealing. In this light, the emerging informal agenda for the Cancún meeting suggests an opportunity with which the North-South conflict has not been presented for many years.

Again it should be noted that this convergence of views concerning an appropriate North-South *agenda* and an appropriate *process* of ongoing negotiation is far more important that the product of any particular meeting, such as the one now planned for Cancún. If the convergence reflects changing attitudes about the benefits to be achieved from breaking the present North-South deadlock, then progress, or a lack of it at any particular moment in time, will be far less crucial than it has been in the past. Without this process of cognitive convergence, random events substantively unrelated to the North-South dilemma can have a major impact on the "success" or "failure" of individual meetings; with convergence, random events can only temporarily slow down or speed up a process which will essentially be driven by autonomous forces.

Any optimism, however, must be balances with a great deal of caution. Both the history of the conflict and the record of the symposium illuminate the barriers facing a constructive negotiating effort, no matter how carefully agendas are drawn, interests are balanced, and linkages are disguised to avoid stalemate.

In the first place, it is surely far from clear that "cognitive evolution" within the North and South relating to mutual interests and conceptions of international economic and political/institutional relationships has yet produced the degrees of convergence required to begin what will inevitably be a long process of negotiations. Outside the symposium, the failure to agree on a modus operandi to launch the "Global Negotiation" suggests limited North-South convergence regarding definitions of problems or the belief that differing goals will yet allow room for positive-sum bargaining. And within the symposium, the clashing perspectives of Rostow and Weintraub on the one hand and Haq, Helleiner, Mills, and Streeten on the other underscored the distance yet to be traveled. The severe differences over such issues as new versus old agendas, regional versus global approaches, linked verses de-linked, functionally specific negotiations, and short-term versus long-term emphases, which arose again and again in the symposium, were not a reflection of participant idiosyncracies; they very accurately reflect fundamental disagreements on values, goals, strategies, and tactics which exist wherever the North-South conflict is analyzed, debated, or negotiated. And so long as the disagreements remain as strong as those revealed in the symposium, optimism regarding short-term movement toward

North-South accommodation is unfortunately misplaced.

This continuing skepticism is illustrated by one particularly revealing exchange in the symposium. Donald Mills closed Session I with a prediction that the South would never relinquish its insistence on the "global" nature of the negotiations and on long-term "structural reform." Walt Rostow responded to this statement by offering the following compromise: let talks on the needs for such global reform continue in the halls of the UN while states immediately push ahead at the regional level to reach agreements on responses to short-term, functionally specific problems.[25] While Rostow's gesture to Donald Mills at the personal level represented a genuine attempt to accommodate, it was also indicative of a perspective which guarantees continued stalemate. For the South has already passed judgment on Rostow's suggested division of labor. It has witnessed eight years of Northern attempts to talk issues to death at the "global level" in the UN forums, while diverting all "action issues" to specialized agencies (e.g., the Bank and the Fund) and bilateral frameworks where the structural environment precludes outcomes responsive to Southern concerns.[26] From the Mills (Southern) perspective, therefore, acceptance of the Rostow division of labor would imply an acceptance of total defeat in pressing for long-term structural reforms in the present international system.

Continued fundamental disagreements aside, there is a further question concerning the willingness and/or capacity of the United States to give North-South issues the salience required to produce serious thought, let alone possible resources. As Joan Spero noted, it is less than certain that the United States can give *short-term* issues related to the North-South conflict the attention they deserve. This theme was elaborated in some detail by Ed Hamilton in his Session IV presentation, which stressed the *derivative nature* of United States policy toward the South at almost all times since 1945, a tendency which he and most observers suspect will be even stronger in the early years of the 1980s, given the domestic and foreign priorities of the Reagan administration.[27]

Faced with these barriers to the process of North-South accommodation of any significance, neither classical realists nor their neorealist successors will assign more than negligible probabilities to the prospects for progress in North-South relations in the first half of the new decade. But it may be worth pondering the tentative decision of most Northern states to accept invitations to the Cancún summit months before American participation has been confirmed. It is also worth noting that Great Britain and West Germany sided with the

United States in opposing the compromise UN resolution which would have allowed the "Global Negotiation" to begin only to avoid an isolated United States veto; the two countries were prepared to support the resolution and changed their position only when it became clear that the United States was adamant in its opposition.

This willingness on the part of most other Northern states to begin negotiations anew can be interpreted in one of two ways. First, it may be seen as a cynical willingness to keep talking without any thought of serious negotiations. Alternatively, it may be interpreted as a sign that a process of evolution is taking place in many Northern capitals regarding the analysis of the costs and benefits involved in continued stalemate, particularly in the face of the broad range of novel problems unfolding in the coming decade.

If the latter interpretation is closer to the truth, then there is some reason for cautious optimism which the victory of Francois Mitterand in the recent French presidential election can only strengthen. And there is further reason for it when one recognizes that Southern countries involved in the Cancún planning effort thus far have backed away from all of the NIEO "buzz words" and phrases which have consistently evoked negative Northern responses since 1974. Their understanding of their own problems and their interpretation of the perspectives, desires, and fears of their negotiating opposites in the North would also seem to be evolving in the direction that will be required if serious negotiations are to begin.

Thus, while there is room to conclude with some optimism, it must be tempered by the presence of two crucial uncertainties: first, the uncertainties surrounding the delicate efforts of the Cancún summit to (re)constitute a *process* of constructive dialogue and negotiations; and second, the uncertainties regarding the evolution of the Reagan administration's thinking as it affects the North-South conflict, uncertainties significantly enhanced by the very low saliency of the issue within the administration in its early months.

On this ambivalent note, the reader is urged to peruse the many revealing presentations and exchanges which occurred in the symposium once again. For perhaps he or she can find in those passages the grounds for a more definitive prognosis for North-South relations in the 1980s than that offered in this postscript.

References

[1]For an incisive discussion of the "market imperfections" issue, see Gerald Helleiner, "World Market Imperfections and Developing Countries," in William R. Cline, ed., *Policy Alternatives for a New International Economic Order* (New York: Praeger, 1979).

[2]This point will be expanded below.

[3]For a brief examination of the "structural reform" issue, see Roger D. Hansen, "North-South Policy: What's the Problem?" *Foreign Affairs* 58, no. 5 (Summer 1980), pp. 1104-1128.

[4]It should be noted that among most development economists, Northern and Southern, the "gap" as a policy issue is almost irrelevant. What is of far greater concern to them are the *distributional results* of the past twenty years of growth. They are concerned that many developing countries and large groups of individuals within others have gained so little — if at all — from the growth process "encouraged and assisted" by the present international economic system.

[5]Stephen D. Krasner, "North-South Economic Relations," in Kenneth Oye, ed., *Eagle Entangled* (New York and London: Longman, 1978), p. 141.

[6]Ibid., pp. 141-42.

[7]Several qualifications must be made at once. First, not all analysts covered by these two labels accept all of the postulates criticized below. Second, there is generally a distinct difference between "classical realists" and "neorealists" (a group Ernst Haas has subdivided into "liberals," mercantilists," and "mainstreamers") in the treatment of "power." The more qualified the neorealists become in their definition and use of the concept, the fewer my exceptions to their analysis in general. The Haas distinction will soon appear in a forthcoming special volume of *International Organization* devoted to the concept of international "regimes."

[8]See, for example, Robert W. Tucker, "The Purpose of American Power," *Foreign Affairs* 59, no. 1 (Winter 1980-81), pp. 241-274.

[9]For a very insightful critique of international relations literature's general weaknesses in analyzing power issues from this definitional starting point, see David A. Baldwin, "Power Analysis and World Politics: New Trends Versus Old Tendencies," *World Politics* (January 1979), pp. 161-194, and his more recent contribution, "Interdependence and Power: A Conceptual Analysis," *International Organization* (Autumn 1980), pp. 445-470.

[10]Kenneth Waltz, *Theory of International Politics* (Reading, Mass.: Addison-Wesley, 1979), pp. 191–192.

[11]Ibid., p. 192.

[12]Ibid.

[13]Baldwin, "Power and Analysis and World Politics," p. 163.

[14]Ibid., p. 165.

[15]"New" only in the degree of importance attached to the perspective.

[16]For a most interesting, insightful, and judicious example of this general movement, see Robert O. Keohane and Joseph S. Nye, *Power and Interdependence: World Politics in Transition* (Boston: Little, Brown, 1977).

[17]Krasner, "North-South Economic Relations," p. 140. Krasner concluded that the Southern strategy "virtually precludes compromise between the North and South."

[18]See, for example, Stephen Krasner, "U.S. Commercial and Monetary Policy: Unravelling the Paradox of External Strength and Internal Weakness," *International Organization* 31,(Autumn 1977).

[19]For an illuminating sketch of the growing domestic constraints on policy coherence in either the foreign or domestic realm in the United States which has vital implications for "power" analysis, see Lester C. Thurow, *The Zero-Sum Society* (New York: Basic Books, 1980).

[20]For evidence that this cognitive process is already under way in certain functional areas, see Joseph S. Nye, "Energy and Security," in *Energy and Security* (Cambridge, Mass.: Ballinger, 1981), pp. 3–22.

[21]See Hansen, "North-South Policy: What's the Problem?"

[22]For a path-breaking analysis of the potential role of new knowledge and altering perceptions in negotiations such as the present North-South "dialogue," see Ernst B. Haas, "Why Collaborate? Issue Linkage and International Regimes," *World Politics* (April 1980), pp. 375–405. The term "cognitive evolution" is borrowed from Haas's essay.

[23]For a more comprehensive analysis of the so-called "absolute poverty" problem and the challenges it poses to the North-South conflict, see Roger D. Hansen, *Beyond the North-South Stalemate* (New York: McGraw Hill for the Council on Foreign Relations, 1979), Chapters 8 and 9.

[24]For a most comprehensive and insightful discussion of many of these issues, the reader is referred to Paul Streeten's contribution to a forthcoming collection of essays honoring Sir Arthur Lewis, edited by Andrew S. Downes and published by the World Bank.

[25]To quote Rostow: "You would assign to a global level whatever people want to argue about in the New Economic Order" (Session IV).

[26]For a detailed analysis of this U.S. strategy since the mid-1970s, see Ronald Meltzer, "Restructuring the U.N. System," *International Organization* 32, no. 4 (Autumn 1978), pp. 993-1018.

[27]There is a general agreement that this tendency predominates except in those periods when international trends lead Washington to think of "the South" in the context of the East-West conflict — e.g., the very late 1950s and the early 1960s.

LIBRARY OF DAVIDSON COLLEGE

Books on regular loan may be checked out for **two weeks**. Books must be presented at the Circulation Desk in order to be renewed.

A fine is charged after date due.

Special books are subject to special regulations at the discretion of the library staff.

JAN -5.1987

JAN. 18.1987

FEB. -1.1987

FEB. 15.1987

Library of
Davidson College